Contents

KEY POINTS

- Fishing provides food, income ($US80 billion per year) and employment for 200 million people. However, fishing does have environmental costs which threaten marine ecosystems and the sustainability of the fished resource.

- The greatest impacts of fishing tend to occur when an unfished environment is first exploited. The relative impacts of fishing depend on the balance between fishing and natural disturbance.

- Fishing gears are designed to maximize yields of target species while minimizing the costs of capture. These gears have direct and indirect effects on the marine environment, resulting in by-catches of unwanted species and damage to marine habitats.

- Sand and gravel seabeds in shallow water are relatively resilient to the effects of towed gears, such as trawl nets, because they are adapted to natural disturbance. Habitats most at risk are coral reefs, maerl beds and seagrass meadows. Recovery may take many years, especially in the deep sea.

- Species most at risk from the direct and indirect effects of fishing are characterized by late maturity, large body size and low potential rates of poputation increase. Species least at risk have high population recovery rates and morphologies that can withstand contact with fishing gears.

- The populations most at risk from pelagic gears are those of some seabirds (especially albatrosses), turtles, sharks and marine mammals. By-catches of birds or marine mammals are high in some long-line, gill or seine net fisheries but mitigation measures can reduce by-catches without reducing catches of target species.

- Lost nets and traps may continue to catch fish. Such 'ghost fishing' will continue until the gears are broken up or overgrown with fouling organisms.

- Fishery discards provide important food material for surface-feeding seabirds, whose populations will inevitably suffer if discarding practices are reduced. Discards that sink provide food for benthic scavengers.

- Fishing can have indirect effects on the structure of marine communities and ecosystems. Some coral reefs have shifted from coral- to -algal-dominated phases following fishing. In many other ecosystems, however, fishing does not have clear effects on interactions and changes are largely due to the loss of vulnerable species.

- Fishing as with other human activity that provides food, income and employment, has some undesirable impacts on the environment. Managers have to balance the costs of these impacts against the benefits derived from the food, income and employment that fishing provides and against the costs of producing protein in other ways.

- Effective fisheries conservation will maximize the long-term yields of protein and income from fisheries while minimizing environmental impacts. Marine reserves (no-take zones) can help protect vulnerable habitats and species of conservation concern.

1 Commercial fishing: the wider ecological impacts

> **Issues**
>
> Fisheries provide food, income and employment for 200 million people. Fishing gears are designed to catch edible and marketable fish or shellfish, but they also catch non-target species and damage marine habitats. The direct effects of fishing have indirect effects on the structure and function of marine ecosystems. We need to know the causes and consequences of fishing effects in order to apply valid conservation measures. These will help guarantee long-term yields of food and income from fisheries while minimizing their environmental impact.

1.1 Introduction

"It is impossible in these times to develop a 'natural' ecology, one that ignores the impact of Man. Ecology should inspire a wiser management of nature: the feedback should work"
Ramón Margalef (1968)
Perspectives in Ecological Theory

Commercial fishing gears have been refined over time to maximize the yield of harvested species and minimize the costs and effort expended in their capture. While past endeavours were made to improve the size and quality of the landed catch, little attention was paid to the quantity and fate of incidental and discarded catches of non-target species. Such by-catch species might include bottom-dwellers, like crabs and starfish, or pelagic mammals, turtles, sea snakes and seabirds. Additionally, the extent to which different marine environments might have been damaged by fishing gears was rarely an issue of pressing concern until the early 1990s, when research interest in this topic began to gather momentum. Concerns were first raised about the unseen damage caused by fishing in a petition presented to Parliament in London in 1376. Five hundred years later, the Royal Commission on Trawling (1883-85) recommended that government money be allocated to find out how fishing gears worked and what effects they had on stocks and the sea bed; a century later we are still asking these same questions.

Globally, fisheries are now reaching the limits of exploitation. In many marine ecosystems, 20% or more of primary production is required to support fisheries, and biological processes rather than the power of fishing fleets often limit fish production. As a result, few potentially productive sea areas remain unfished and, despite increased investment in fisheries development, the rate of expansion in the world catch has tailed-off in recent years (Figure 1). Global marine catches are unlikely to exceed substantially the current levels (1995 data) of about 90 million tonnes per year without dramatic changes in management methods. Catches may not even be sustainable at this level.

Global fish catches are approaching their biological limits

1

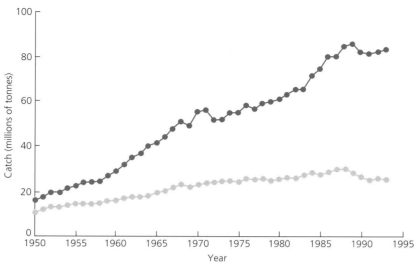

Figure 1 Global marine fish catches since 1950. Filled circles show the catch by all nations and open circles the catches by industrialized nations (data from FAO, 1996).

Many exploited fish stocks now show signs of overfishing and a few, such as the northern Atlantic cod *(Gadus morhua)* stock off Newfoundland, have been fished to economic extinction in recent years. Since marine fish in international waters have been regarded generally as common property, considerable global fleet overcapacity has developed; with too many vessels chasing too few fish. More than one million fishing vessels currently exploit the global oceans; twice as many as in 1970. The result is that although global fisheries currently yield $80 billion annually (as sale value of catches), they cost at least $50 billion in subsidies. According to the FAO, 70% of the world's 80 most important commercial fish stocks are now either fully exploited, overex-

Table 1 The ten species that contributed most to global marine catches in 1995 (data from FAO, 1996).

Common Name	Species	Catch (million tonnes)	% catch
Peruvian anchovy	*Engraulis ringens*	8.64	9.4
Chilean Jack Mackerel	*Trachurus murphyi*	4.96	5.4
Alaska pollock	*Theragra chalcogramma*	4.69	5.1
Atlantic herring	*Clupea harengus*	2.59	2.5
Skipjack tuna	*Katsuwonus pelamis*	1.56	1.7
Chub mackerel	*Scomber japonicus*	1.56	1.7
South American pilchard	*Sardinops sagax*	1.50	1.6
Yesso scallop	*Pecten yessoensis*	1.42	1.5
Atlantic cod	*Gadus morhua*	1.26	1.4
Largehead hairtail	*Trichiurus lepturus*	1.24	1.3

1

ploited, depleted or rebuilding after overfishing. The global fishing industry is based on remarkably few species. Although there are some 13,000 species of marine fish, just 17 each contribute more than 1% of world landings and the top ten account for 32% of the total (Table 1).

Fish provide one of the most significant and potentially sustainable sources of animal protein for human consumption. They account for about 29% of all animal protein consumed in Asia (about 60% in Indonesia), and their exploitation generates income and provides employment for millions of people worldwide. However, if they are to continue to do so in the long term then harvesting practices need to be managed in an environmentally sympathetic and ethical manner. The seas and oceans patently do not contain inexhaustible resources as was asserted by Thomas Henry Huxley to successive British Fisheries Commissions in the 1860s. The notion of sustainable exploitation needs to be widened to include not only fished species but also the ecosystems from which they are extracted.

Although sustainable exploitation is a laudable goal, many fisheries are not exploited sustainably. There are two main reasons for this. First, there is usually common access to fisheries or access is ineffectively controlled. This leads to a 'race to fish' where it is better to catch a fish today because if it is left in the water someone else may catch it tomorrow. Second, current levels of demand for fish protein and income are so high that too many fishers are chasing too few fish. Overfishing has been exacerbated by subsidies that sustain economically inefficient fisheries. In places such as the Philippines, fish-

eries are collapsing following habitat degradation. The reason is socio-economic. Fishers, faced with declining catches, fish more intensively, often with habitat-destructive gears, until the resource collapses. Small-scale fishers targeting coral reefs, for instance, are often poor with few alternative prospects for employment. Solutions to this problem will only come in the short-term from the creation of alternative employment opportunities for fishers and in the long-term from the stabilization of human population growth.

In this booklet we have chosen deliberately to exclude any discussion of the direct effects of fisheries on target organisms, as these issues have been dealt with at length in a number of excellent texts (see 'Further Reading'). Instead, we have focused our attention on the secondary effects of fisheries activities; effects that until recently, have been overlooked by many fisheries managers.

In order to appreciate the effects of fishing and fishing gears, it is necessary to consider the mode of operation of different gears and the interactions that occur both on different ground types and with different assemblages of resident organisms. That accomplished, we then consider the following topics: the impacts of litter generated by fishing gear; modifications to habitats that ensue from fishing activities and the direct and indirect effects of fishing on a range of non-target organisms. This will lead on to an assessment of the mechanisms available for monitoring these effects; and finally we shall describe the implications of our findings for conservation and management decision making.

1

Summary

Fishing gears are designed to maximize yields of target species while minimizing the costs of capture. These gears have direct and indirect effects on the marine environment, resulting in by-catches of unwanted species and damage to marine habitats. With global fish catches approaching their biological limits, fishing effects are compromising the sustainability of some fisheries that provide vital sources of food and income, and are threatening some species and habitats of conservation concern.

2 Fishing Gears and their Operation

2.1 Gear selectivity

An ideal fishing gear is 100% efficient in capturing the target species and no other. That is, it captures every target organism over a certain size within its operating area, while excluding all others. Needless to say this ideal is rarely met. The closest approximation derives from the pelagic realm. Here, in the industrial fishery for species such as Peruvian anchovy *(Engraulis ringens)*, largely monospecific shoals of fish may be captured with encircling nets. Traditional spearfishing also generates no by-catch, neither do specialized harpoon fisheries for tuna. Off Cape Cod, USA, for instance, huge bluefin tuna *(Thunnus thynnus)* are fished for by electrified harpooning, after tracking by spotter planes; a technique that readily becomes cost-effective since individual tuna may fetch more than $30,000 dockside in Japan. The Western Pacific pole (rod) and line fishery for tuna limits by-catch to less than 1% of total catch. Such examples, however, represent the exceptions rather than the rule.

While all fishers aim to maximize their mar-

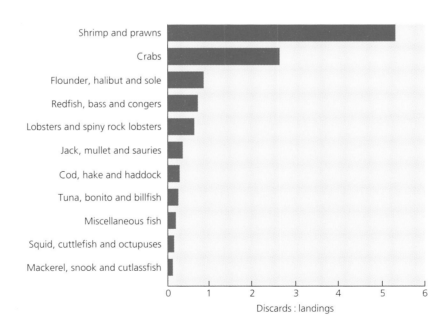

Figure 2 The ratio of discarded bycatch to landings (by weight) in 11 major fisheries.

2

ketable catch, most gears capture varying proportions (sometimes >90%) of unwanted material together with unsaleable undersized individuals of the target species (Figure 2). Where it is unlandable, this portion of the by-catch is discarded at sea, amounting presently to some 27 million tonnes of trash biomass discarded annually worldwide. Discarding may be a side-effect of management systems intended to regulate fisheries. For instance, non-transferable quotas may cause discarding of over-quota catch and species-specific licensing may cause discard of non-licence, but still commercially valuable, species. In the Northwest Atlantic, in an attempt to avoid discarding of by-catches, an aggregate quota for northern Atlantic cod, haddock (*Melanogrammus aeglefinus*) and pollack (*Pollachius pollachius*) was introduced in 1989. The discard rates of sublegal sizes of particular species will also vary between-years with variations in their abundance.

Put into context, discarding represents about 25% of the world's capture fisheries, or more than four times the entire catch of United States fishers. In the southern North Sea 56% of fish captured are discarded by the beam trawl fishery, and 475,000t of fish, offal and benthic invertebrates are discarded into the North Sea annually, 50% of which is consumed by seabirds, the remainder by bottom-dwelling predators and scavengers (Figure 3). Shrimp fisheries generate the highest by-catches worldwide. Some 10kg of undersized finfish are captured in shrimp trawls in the Gulf of Mexico for every 1kg of shrimp retained. Off the Brazilian coast, discards from the trawl and shrimp fisheries are comparable to the size of the total landed catch. The trawl fishery for prawns in Northern Australia catches some 200 species of benthos, but target prawn species account for less than 11% of the total catch weight. In 1990, it was estimated that the Japanese fishing fleet killed 41 million non-target sea creatures (including 39 million other fish, 700,000 sharks, 270,000 sea birds, 26,000 cetaceans and

Figure 3 When marketable fish such as these cod and sole are caught, large by-catches of unwanted fish and invertebrates are also taken (Photograph: © S.Jennings).

400 sea turtles). By-catch mortality is now the main source of shark mortality worldwide and some shark species are now considered to be endangered. The Hawaiian longline fishery for yellowfin and bigeye tunas *(Thunnus albacares* and *Parathunnus obesus* respectively) and swordfish *(Xiphias gladius)* catches over 100,000 sharks annually, more than 50% of which die. Eighty percent of juvenile billfish (ie sailfish and marlin species) die when discarded from longline fisheries in the Gulf of Mexico and off the East coast of Florida, USA. For most fishers, discarding is accepted practice but the Norwegian Government permits no discarding of fish at sea in its waters. At the other extreme, all species may be landed in some low catch-rate multi-species fisheries in the Mediterranean Sea.

Discarding by-catch is a highly visible sign of the inefficiency of many fisheries. It is in the best interests of fishers to reduce by-catch; hooks which snag seabirds clearly reduce the capacity of baited longlines to capture target fish. The aggregate economic losses through discarding in many fisheries and regions of the world could easily approach the value of the landed catches. This is especially true in crab fisheries. Discarding depletes stocks, reduces spawning potential and results ultimately in diminished yields.

Since smaller fish generally attract lower market prices, skippers are prompted to fish and discard catches and fish again in search of shoals of larger, higher value fish. Such 'high-grading' represents an additional source of discard mortality, the scale and impact of which is very difficult to quantify. In the Philippines and elsewhere in southeast Asia tiny fish and even fish larvae are in demand, being used in the manufacture of fish sauces.

A catch dominated by unwanted species is economically inefficient in terms of the time needed to sort it, and fish that are handled but generate no revenue are becoming a larger fraction of the total catch. The development of economic uses for such species is one obvious way to reduce such wastage. Discarding is thus gradually becoming a less and less politically acceptable option, especially when it coincides with a growing demand for fish and fish products generated by healthy-eating promotions. Discarding fishery wastes also results in the unnatural cycling of large amounts of biomass from the sea bed to the surface, affecting the feeding behaviour of some consumers at the surface, in mid-water and on the sea bed. The FAO predicts that through better fisheries management and reduction of waste it would be feasible to add another 10-20 million tonnes per year of fish for direct human consumption. Discarding would be reduced by improving gear selectivity and improving the interactions between fishers, fishery scientists and policy makers.

In the past, developments in gear design have generally been made with a view to creating bigger, more selective harvesting devices. An important practical goal has always been to create a gear that can withstand the physical rigours of prolonged use in the marine environment, particularly when that involves contact with the sea bed. Now, we need to seek out ways in which fishing practices can be designed to create as little damage to the environment as possible, compatible with the need for continued sustainable exploitation of fishable stocks.

There is another side to this coin though. Inadvertent by-catches, particularly of large marine mammals, sharks, turtles or spiny spider crabs *(Maia squinado)*, can damage fishing nets and

Catches dominated by unwanted species cost time and money to sort

Unwanted by-catch can damage fishing gears

traps, incurring high repair costs and loss of fishing opportunity for fishers. Available figures suggest that 2-10% of gear is so damaged on an annual basis. The entanglement of sperm whales *(Physeter macrocephalus)* in longlines results in the loss of major sections of fishing gear as well as the catch. Unwanted taints may also remain after contacts with non-target species; for example, the black epidermal secretions of the basking shark *(Cetorhinus maximus)* allegedly ruin netting that has rubbed against it. Scottish salmon fishers claim that they can smell when a seal has been in their set nets and that this odour can deter fish from entering the net for several days. Non gear-damaging interference by non-target species can also substantially reduce the catch rate of target species. Bait losses of up to 70% have been reported due to seabirds interfering with longlines.

The following description of how fishing gears operate is not intended to be encyclopaedic in scope. For such, readers are referred to standard texts, a selection of which are listed in 'Further Reading'. Fishing methods can be split broadly into two categories: passive techniques which involve the use of set gear, such

Drift nets and long lines are used to catch pelagic fishes

as static or drifting nets, hooks or traps; or active techniques such as trawling and seining in which the target organisms are pursued.

2.2 Pelagic drifting gears

Monofilament drift nets have been the subject of much media attention. These so-called 'walls of death' were once deployed in fleets >10km long to capture pelagic tuna and squid on the high seas (Figure 4a). Oceanic squids have the greatest potential for fishery development, and drift-netting for squid is a relatively recent development; initiated by the Japanese in 1978. Salmon are intercepted in the Pacific by squid drift nets. In the North Pacific, an estimated 170,000km of gill nets are available to the major fisheries, and the interception of migratory salmon by these nets is a recognized problem. A typical Taiwanese vessel might deploy about 20km of 10m deep monofilament gillnet to drift free overnight, every night; the fleet deploying some 14,400km of net each night during the May to September squid-fishing season. These nets, that are invisible to most marine organisms (and SCUBA divers), inadvertently entangle dolphins and other cetaceans, turtles, and diving birds. In 1991, environmental concerns about these

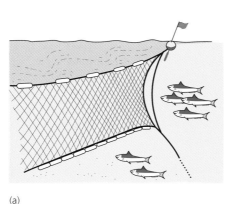

(a)

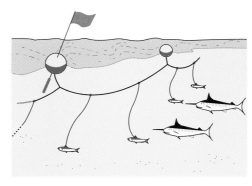

(b)

Figure 4 Pelagic drift net (a) and long-line (b).

2

huge nets resulted in a UN moratorium on nets longer than 2.5km. However, the illegal use of nets exceeding these legal length limits continues, since enforcement is difficult.

Pre-dating this UN ruling, the Australian fisheries authorities introduced their own 2.5km maximum length for pelagic drift nets in 1986. This so reduced the operations of the gill net vessels off northern Australia, that the fishery became uneconomic and fishing ceased, eliminating the dolphin by-catch in a fishery where 14,000 dolphins had been caught between 1981-1985. In 1994, the EC proposed new legislation to ban high-seas drift nets of any length by the end of 1997, but agreement has not been reached as of early 1999. Another problem with gillnets is the high drop-out rate of loosely ensnared dead or dying fish during net retrieval.

More selective tuna- and swordfish-capture strategies use drifting longlines tens of kilometres long (Figure 4b), though these have their own environmental side-effects, particularly in terms of unwanted catch of juveniles, sharks, turtles and seabirds (especially albatrosses and petrels). The central line is of heavy breaking strain, and carries several thousand hooks baited with squid or mackerel on multiple branch lines. Chemical light sticks may also be attached to the lines to act as lures; one vessel maybe using 5000 light sticks per trip. Setting a minimum hook size to avoid particular species or juveniles of target species is the usual conservation measure adopted in long-line fisheries for fin-fish, but choice of hook is a compromise since a mixture of species may be targeted. Changing hook sizes within the commercial range does not seem to have much effect on by-catches of seabirds.

In the CCAMLR (Convention on the Conservation of Antarctic Marine Living Resources)

area of the Southern Ocean, longline fisheries were introduced in the mid-1980s to catch Patagonian toothfish *(Dissostichus eleginoides)*. They have expanded rapidly in recent years and a considerable amount of this fishing is currently illegal and/or unregulated. A variety of methods are being used to scare or distract seabirds away from the most vulnerable area, where the lines are set. Acoustic scarers, towed buoys and magnetic deterrents are largely ineffective. Limited short-term deterrence has been achieved using water hoses. The use of offal lures to tempt birds away from the lines during setting has been more successful as a palliative measure. In the long-term, however, offal discharge (even away from the haul) simply increases the numbers of seabirds associating with fishing vessels, putting more at risk. The most widespread and effective measures are poles (called tori poles) with attached lines bearing scarers (streamers on swivels) which move unpredictably in the vicinity of the setting area. Additionally, it may be possible to dye bait to make it less visible to seabirds, and the development of smart hooks (that retract the point until a safe depth underwater has been reached) is being pursued.

Several methods are being used to reduce the time that baited hooks stay near the surface and are thereby available to catch and kill seabird species feeding predominantly at the surface. These include the use of thawed bait which is less buoyant than frozen bait, machines to cast bait beyond the turbulent area in the wake and weights on branch lines to sink bait more quickly. Aerial lines with streamers flapping above towed longlines were developed originally by Japanese fishers and reduce seabird by-catch by 30-75%. No single measure is suitable for every fishery or vessel and each longline fishery needs to determine the most effective measures to suit its purpose. Indeed, some measures in some areas and/or circumstances may

2

actually increase the catch of other non-target species. Much current effort is being devoted to devising appropriate mitigating measures for all regulated longline fisheries and areas.

For seabirds like albatrosses that are mainly diurnal, restricting setting to the hours of darkness is highly effective in reducing by-catch and can virtually eliminate it, except on moonlit nights. Reducing deck lighting is an additional refinement safeguarding seabirds, if not fishers, from hitting vessel superstructures. However, other seabirds, especially petrels and notably the white-chinned petrel *(Procellaria aequinoctialis)*, are adept at fishing at night and significant by-catch of such species can result.

The use of mitigating measures (especially streamer lines, weights and thawed bait ideally with punctured swimbladders) has become compulsory in several fisheries since CCAMLR introduced the first mandatory provisions of this kind for the whole of the Southern Ocean in 1991. The use of night setting is also compulsory within the CCAMLR area, but in fisheries where commercially viable catch rates require more than one set per day there is considerable resistance to introducing such measures. There are obviously problems with using night-time setting in high latitudes, due to insufficient hours of darkness in some seasons. CCAMLR currently resolves this by closing the entire fishery during the relevant summer months.

In the longer term, techniques for setting longlines under water will be a more effective means for avoiding incidental mortality of seabirds. A variety of underwater setting devices are currently being developed. Even when these become successful, however, it will take some time for them to become the industry standard and methods may not be applicable to all existing vessels. Changes in vessel trim when vari-

ously laden and under different sea conditions may affect the efficacy of these devices. A setting funnel, through which the baited longline passes into the water astern of the vessel, used in conjunction with aerial streamers, has recently been advocated as the best method for reducing seabird by-catch, principally of fulmars *(Fulmarus glacialis)*, in the North Atlantic.

2.3 Bottom set gears

Usually, static bottom gears are anchored to the sea bed and left to fish passively. They either entangle target species that wander into them (gill or trammel / tangle nets), or attract target species to them (baited hooks, or traps, pots, creels) (Figure 5). By and large, static gears have less impact on the environment than towed gears (see below). That said, however, damage to coral reefs can be sustained during hauling of set nets, and this problem has been exacerbated by increasing mechanization, like the introduction of power blocks on fishing vessels. Large-mesh monofilament nets set on the sea bed, mainly for bottom-foraging fish, pose a threat to harbour porpoises *(Phocoena phocoena)* throughout their range and especially in the North Atlantic.

Benthic longlining is a selective method of catching bottom-foraging fish, like cod, halibut *(Hippoglossus hippoglossus)*, some tunas and sharks. Hook size can be increased so as to avoid capture of unwanted species and juveniles of target species with small gapes. Different baits have appreciably different abilities to remain intact on the hooks. For example, squid or defrosted cuttlefish maintains its integrity on the hook. Bait freshness is important in preventing it falling off the hooks.

The type of bait is important too, since particular baits are attractive to different target species. Generally speaking, targeted species are less

But mitigation measures can be used to reduce by-catch

Bottom-set nets, lines and traps catch species living on or close to the seabed

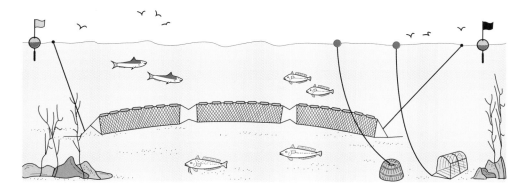

Figure 5 Traps, pots and set gill nets used to catch species living close to the sea bed.

attracted to baits of the same, or closely related, organisms, so crabs are better potted for using fish than dead crabs as bait. Deployment time also has important implications for catch rate. Usually, baited traps set for cephalopods *(Nautilus,* octopods) require lengthy soak periods.

Since set (net and pot) fisheries are static and more selective, the areas affected by such gears are likely to be insignificant compared with the widespread effects of mobile gear. Thus, set gears are often more suitable for sustainably exploiting areas of high conservation value. In the Bering Sea, for instance, pot fisheries intercepted far fewer species than did bottom trawl fisheries, because such gear was more selective. Since the late 1800s large fixed traps have been used to catch North Atlantic cod in Newfoundland and Labrador. Even though this technique avoids the damage associated with trawling, significant by-catch problems can still arise. The salmon by-catch problem from these traps is rectified by requiring their return to the water. Considerable work was also needed to discourage whales from approaching the traps.

2.4 Pelagic mobile gears

Gears like mid-water trawls and purse seines are used to capture free-swimming (pelagic) fish in the water column (Figure 6). Both these gears can be huge, capable of taking perhaps 400t of catch per single deployment. The Icelandic Gloria 'jumbo' trawl is 2,000m long and 110 x 170m at the mouth. It is designed to catch Icelandic redfish *(Sebastes marinus)*, a non-shoaling fish that lives in deep water on the mid-Atlantic ridge. Purse seine fishers may use spotter planes as well as sonar to locate fish shoals around which the nets are deployed. The circle enclosed by some purse seines can have a diameter of over 600m and the base of the net is drawn closed to enclose the shoal. They are used to catch fish like tunas, salmon, herring *(Clupea harengus)* or mackerel *(Scomber scombrus)*, swimming near the sea surface in the warmer, highly oxygenated water above the thermocline. Tunas may be chased by speedboats or corralled using explosives before net encirclement.

Fish aggregation devices (FADs) simplify the catching of fish in warmer waters by taking

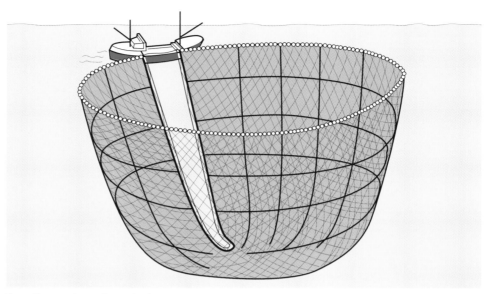

Figure 6 A purse seine used to catch pelagic species.

advantage of their tendency to aggregate around floating debris. Artificial FADs can be made by hanging tree branches or plastic tents beneath rafts, and the aggregated fish are then harvested by trolling lures or setting the gear (purse seine, baited hooks or gillnet) close to these objects.

Secondary impacts on some of the larger gelatinous zooplanktonic organisms (salps, medusae, ctenophores) as a result of collisions with nets are inevitable but, to our knowledge, this topic has never received any scientific attention. The commercial fishery for krill *(Euphausia superba)* off South Georgia takes small numbers of young fish, including pre-recruit stages of the important mackerel icefish *(Champsocephalus gunnari)*, but takes almost no by-catch of seabirds and marine mammals. Impacts on such non-target nektonic organisms as sea mammals, turtles and sharks are important in some purse-seine fisheries but can be mitigated by appropriate modifications to gear and fishing technique, as will be seen later

The presence of salps and jellyfish as by-catches in pelagic gear certainly hinders industrial krill processing, and the handling of fishing gear contaminated with cnidarian tissues can result in some fishers developing skin allergies (dermatitis) after sensitization brought about by repeated exposure to nematocyst toxins.

2.5 Bottom mobile gears

Bottom-fishing, or demersal, towed gears are designed to catch target species living close to, or on, the sea bed (Figure 7). Beam trawls were developed to catch shrimps *(Crangon crangon)* or flatfish, such as plaice *(Pleuronectes platessa)* and sole *(Solea* spp.). Early versions, which used a wooden beam to extend the mouth of the net between a pair of skids (shoes), were relatively small and lightweight. As bigger, more powerful vessels became available, the size and weight of beam trawls increased as they incorporated stronger beams and heavier chains and shoes. Some fishers have added rollers to the beam shoes, to cut down drag and thereby reduce fuel costs. Typically, beam trawls are

Beam and otter trawls catch species living close to the sea bed

2

towed in pairs, one on either side of the vessel.

Some European beam trawlers tow a 12m wide beam trawl (weighing up to 7.5 tonnes in air) on each side of the vessel. The tickler chains (up to 17 on a 12m beam trawl) each dig successively deeper into the fluidized substratum, or a heavy chain mat precedes the net bag, and disturbs buried target species, like shrimps or soles. Chain mats penetrate surface sediments less deeply than tickler chains.

In an otter trawl, the mouth of the net is kept open by two solid boards, functioning like paravanes. These are called otter boards (or doors) and rigged to pull the wings of the net open when towed. In addition, the sediment clouds raised by the passage of the doors help to herd fishes towards the mouth of the gear. Otter boards can be extremely heavy (maybe several tonnes in air) and plough deep parallel furrows in the sea bed as they progress, displacing infaunal animals. Their impact will be continuous only on smooth bottoms. Since they are pulled at an oblique angle to the direction of travel, however, only a proportion of their width gouges the sea bed. Together with developments like steam trawlers and power winches, this gear gradually opened-up access to North Sea waters deeper than 100m at the end of the nineteenth century.

The largest otter trawls (>100m mouth, with nets 0.5km long) are now fished by factory freezer-trawlers. Somewhat smaller trawls are routinely towed between two catching vessels (pair trawling) in European seas. Twin trawl nets linked side-by-side and pulled by one powerful vessel, and requiring only one set of doors, can double the catch whilst incurring only a 30% increase in fuel costs. Demersal gears like these disturb large areas of sea bed. Large beam trawls, however, have the most extensive impact on the bottom since their chains disturb sediment to twice the depth affected by an otter trawl's ground tackle.

On Georges Bank, New England some 3% of the ground was reported to be covered by trawl marks in 1973. More recent estimates from different areas have generated higher values (ca 18%). Intensively fished parts of the North Sea are swept by trawls several times per year. Estimates vary, but figures of 3-5 times per year are often quoted. That may be the average condition. Some areas, however, will be fished more heavily than that, while other, less productive or gear-damaging, areas (e.g. near wreck sites or other snagging points) will remain relatively undisturbed. Thus some 3 x 3 nautical mile sectors of the North Sea are visited >400 times per year, while others are hardly ever fished.

Trawls cause direct damage to the sea bed and its associated fauna

Figure 7 Otter (a) and beam trawls (b) are used to catch bottom dwelling species.

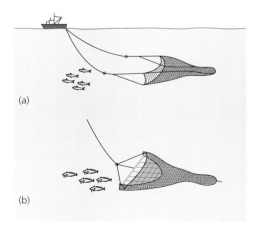

(a)

(b)

13

2

Towed dredges are used to catch shellfish

Fishing with explosives can reduce coral reefs to rubble

Otter trawls are not able to catch demersal bivalve molluscs effectively. For this purpose the towed scallop dredge was developed (Figure 8). Such gear is also used to harvest sea cucumbers off the coast of Maine, USA. Early scallop dredges consisted of a bag that was dragged along the sea bed behind a metal frame, and had fixed-teeth across the bottom bar at the dredge mouth. The design was improved by using spring-loaded teeth that reduced bottom snagging and so could be used over rough ground. The size of the teeth and the width of the belly rings assist size selection of the animals captured. In some cases, several dredges are attached to a wheeled rigid axle in groups of 3 or 4. As many as 20 dredges may be towed simultaneously on each side of a fishing vessel. The great weight and strength of the gear creates considerable disturbance to the ground over which it is towed, overturning rocks and dislodging and crushing many organisms in its path.

In the Adriatic Sea, lightweight dredges, known as rapido trawls because they can be towed at speeds of up to 10-15km h^{-1} (6-8 knots), are used. These are towed three or four at a time, over rather homogeneous sandy grounds, and used to catch shellfish and flatfish. The rapido trawls have a fixed-tooth bar that rakes the sea bed. This clogs as the tow proceeds because large semi-buried bivalves like fan mussels *(Atrina fragilis)* and soft-bodied tunicates become impaled on the teeth.

In parts of the Mediterranean, such as the Alboran Sea, a very destructive towed gear called the coral bar is used to scrape red coral *(Corallium rubrum)* from depths of 30-130m. The gear consists of a heavy bar or cross, of metal or wood, to which skeins of old netting are fixed. It breaks off the corals which then get entangled in the netting allowing retrieval. Greek fishers also use a scraping gear to gather sponges from hard grounds.

Hydraulic dredges either use jets of water to disturb the ground in front of a towed dredge to capture infaunal bivalves, like razorshells *(Ensis* spp.) and cockles *(Cerastoderma edule)*, or use a pump to suck bottom sediments on board ship where bivalves are screened out and the spoil discharged back to sea. In mobile sediments, these dredging techniques are such effective ways of collecting bivalves and lugworms *(Arenicola marina)*, the latter for bait, that devastation of local populations of target species can result, causing knock-on effects elsewhere in the food chain. Wholesale mechanical removal of the top 30cm of sedi-

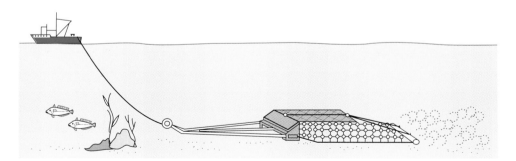

Figure 8 A scallop dredge as used in fisheries off the north-eastern coast of the United States.

2

ment to gain access to razorshells will affect associated infaunal species. Such techniques are strictly regulated for conservation reasons in areas like the Dutch Wadden Sea or the Traeth Lafan Local Nature Reserve in Wales where dredging impinges on the food resources for oystercatchers *(Haematopus ostralegus)*. Traditional methods of harvesting cockles by hand, are less destructive than either dredging at high tide or tractor harvesting at low tide (which latter can result in 100% exploitation of the beach). Other ancillary effects, such as the winnowing of fine particles from the spoil plume and siltation impacting beyond the fished area, may also be important knock-on effects.

2.6 SCUBA diving, explosives and poisons

Hand-collection by SCUBA divers is potentially the most selective to target species, and least damaging fishing method to non-target species. This technique is used to collect surface-dwelling shellfish like scallops, lobsters or abalones, or sand-burrowing razorshells. By and large, diver-based collecting tkes place on a much smaller artisanal scale than with indirect methods of exploitation. However, instances have been reported where wholesale habitat destruction has accompanied SCUBA fisheries, for example for the rock-boring date mussel *(Lithophaga lithophaga)* along the Italian Apulian coast. Here, the illegal method of acquisition can involve the demolition of the rocky substratum with pneumatic air tools. Collection of animals for the souvenir/curio trade (e.g. sea urchin lampshades, sea-horse brooches, conch shells) or for jewellery manufacture (e.g. using red coral; or black coral, *Antipathes* sp. for mourning jewellery) has resulted in local extirpations in certain parts of

Explosive and poison fishing are often banned but enforcement is difficult

the world, with unknown ecological consequences.

The use of explosives (dynamite) and poisons (like sodium cyanide, bleach, DDT) to extract fishes from tropical coral reef or mangrove areas is indiscriminate in that it affects both target and non-target species. Blast fishers often manufacture their fishing bombs from mining explosives or armaments. They fix short fuses to the bombs in an effort to ensure that they explode midwater. A few square metres of coral can be damaged by any single explosion, but the shock wave kills most fishes, as well as some invertebrates, within a radius of 50m or more. Most of these fish are never collected by the fishers. Reefs in some parts of southeast Asia are so intensively blast-fished that areas of living coral have been reduced to rubble and little, or no, reef habitat remains. Heavily dynamited reefs suffer marked reductions in diversity.

Sodium cyanide has been adopted as a 'modern alternative' to traditional fishery poisons made in small quantities from plants such as Derris. Cyanide is used to catch live reef fishes for the aquarium trade and for human consumption. Fish poisoned with cyanide will usually recover, at least temporarily, if placed in clean water. The WWF has estimated that over 6,000 cyanide divers use 150,000kg of poison that affects 33 million coral heads annually. In an attempt to discourage such practices, the Philippines government now test fish tissues for cyanide nationwide to police a cyanide-free fishing policy. However, while species like the humphead or Napolean wrasse *(Cheilinus undulatus)* continue to be considered as gourmet delicacies in Southeast Asia, fetching prices of $187 per kg, then cyanide fishing will remain a temptation.

2

Fishing with poisons, explosives and other habitat-destructive methods like drive-netting (muro-ami or kayakas) is now illegal in many countries. However, enforcing laws preventing such practices is another matter. Intimidation by dynamiters as they tried to prevent police action has been reported recently in Tanzania. The prevention of dynamite fishing at sea depends as much on adequate systems of accountability and security for dynamite used in quarrying and road building on land as on conscientious coastal patrolling.

Summary

Fishing gears catch target and by-catch species. More of the catch is often discarded than retained, and sorting the catch costs time and money. By-catches of birds or marine mammals are high in some long-line, gill or seine net fisheries but mitigation measures can reduce by-catches without reducing catches of target species. Bottom trawls, dredges and explosives have the greatest direct impacts on habitats

3 Effects of litter from fishing gear

3.1 'Ghost' fishing

The non-biodegradability of the modern synthetic fibres used in the manufacture of fishing nets means that snagged and lost trawls, fleets of creels, or torn fragments of mesh floating at sea, may continue to fish for months or even years. This phenomenon is termed 'ghost' fishing. Despite substantial losses of gear, there are considerable practical difficulties in researching the impacts of 'ghost' fishing. It is estimated that 7,000km (20-30% of total set each day) of drift nets were lost per year in a North Pacific fishery and this is particularly hazardous to seabirds. In 1978, for instance, 99 seabirds were recovered from a single 1.5km gill net retrieved at sea south of the Aleutian islands and, in 1981, over 350 dead seabirds were counted during retrieval of a 15km drift net found floating in the central North Pacific. Summer surveys of the incidental catch of marine birds and mammals in fishing nets around the coast of Newfoundland indicated that over 100,000 animals were killed in drifting nets during a 4-year period (1981-1984). It is quite possible, of course, that during their voyaging such nets would have captured many more organisms (birds, turtles, mammals) that had simply decomposed and fallen away unrecorded or been consumed by scavengers.

Lost nets ball-up after a few days or weeks reducing their catching capacity but they may continue to fish at reduced efficiency for many years. In the USA, it has been estimated that $250 million in marketable lobster resources is lost each year to 'ghost' fishing. Recent studies of nets snagged on the sea bed have shown that a typical pattern of captures is observed. Over the first few days, catches decline almost exponentially as the increasing weight of the catch causes the net to collapse. Then, for the next few weeks, the decay of captured animals attracts a large number of scavenging crustaceans. This cycle of capture, decay and attraction continues for as long as the net retains entanglement properties. Once on the bottom, multifilament nets remain tangled, while monofilament nets may, once clear of fish remains and crabs, disentangle, return to an upright position and resume fishing.

The longevity of 'ghost'-fishing nets in the sea will depend upon environmental conditions; storms and swells rapidly abrade and destroy netting, especially when in contact with sand or rock. Nets lost in shallow, clear water become overgrown rapidly with fouling organisms which increases their visibility and reduces their fishing capabilities. Pots (creels, traps) which are constructed of more durable materials tend to 'ghost' fish for longer periods than nets. A natural rebaiting cycle (after scavenger attraction, entrapment and death) in lost pots suggests that an intact pot made of indestructible materials would continue to fish indefinitely. In order to reduce such losses, escape panels are now fitted to many pots used in North America and biodegradable materials are being introduced to reduce losses from 'ghost' fishing.

Litter from fishing gear also affects humans, not only indirectly through the degradation of the aesthetic quality of shorelines, but also

Lost nets and traps that continue to catch are said to be ghost fishing

The longevity of 'ghost' fishing nets depends on environmental stability

3

directly by propeller entanglement, clogging of engine cooling-water intakes, fouling of gill nets and hooks and even injuries to fishers. Caught-up debris also makes nets more detectable to fish, reducing their efficiency.

3.2 Ingestion of fragments of fishing gear by marine mammals, turtles and seabirds

More than 100,000 mammals and seabirds die from eating or becoming entangled in plastic debris each year, including netting, plastic fishing line, packing bands and styrofoam. Compliance by shipping with the MARPOL Convention should reduce this problem, in particular the cutting of loop-forming materials (like polypropylene bands securing bait boxes).

Ingestion of debris, however, represents a lesser problem than entanglement. Most plastic ingested by seabirds at sea is not derived from fishing and, despite high loads of ingested plastic fragments in some species (especially petrels), evidence of mortality resulting from this is surprisingly difficult to find. The loss of chemical light sticks from drift longline fisheries is not uncommon and these may represent a mortal threat to marine animals ingesting them. Estimates have suggested that from one third to one half of all sea turtles have ingested plastic products or byproducts, but few of these are associated with fishing activities. The main threat to turtles is through incidental capture in shrimp fisheries (Section 5.3).

Ingestion of lost fishing gear is rare and most ingested plastic comes from other sources

Summary

Lost nest and traps may continue to catch fish. Such ghost fishing will continue until the gears are broken up or overgrown with fouling organisms. Fragments of lost gear can be ingested by marine animals, turtles and birds, but most plastic in the marine environment comes from other sources

4 Vulnerability of different marine habitats

Maximum fishing effort will be centred in areas of highest productivity and such areas are not equally or equitably distributed. Thus the shallow North Sea produces 3% of global fisheries yield yet covers only 0.004% of global sea area. Most bottom fishing takes place at depths of less than 500m. A country's demersal fisheries resources depend fundamentally on the extent of its continental shelf and, in particular off the western coasts of South America and Africa, proximity to nutrient-rich upwellings. These coastal upwelling regions produce half of the world's fish catch.

Habitats have differential vulnerability to fishing

Compared with the deep sea, however, productive, shallow-water environments are more vulnerable to the effects of man (like nutrient enrichment) and natural perturbations (climate variation, weather effects), and in some areas these processes play an important part in structuring benthic communities. That said, the last two decades have witnessed the development of 'boom-and-bust' fisheries in the deep sea, as pressures to find alternatives to overfished stocks in shallower waters have mounted.

Stable muds are more susceptible to the impacts of fishing then mobile sands

Since fishing is practised on a global scale, the exploitation of high-seas pelagic resources is an international industry which impacts biological resources beyond the jurisdiction of individual countries (Section 7.2). The physical habitat is not at risk here, although some non-target organisms are (Section 5). The vulnerability of any particular habitat will depend on the relative magnitude of natural and anthropogenic disturbances. In this section we compare and contrast the vulnerability of a number of benthic habitats to fishing.

4.1 Mud

Soft muddy grounds generally support the lowest population densities of surface dwellers. However, certain fragile organisms like sea pens *(Pennatula, Virgularia, Funiculina)* live there and are vulnerable to repeated damage by fishing gear. Although species of *Pennatula* are retractile and may withdraw into the mud before being contacted by the gear (whether landed on by pots or impacted by towed gear), it is likely that the frequency and intensity of physical contact is the most important determinant of their mortality. Most mud-dwelling fauna lives within the mud (infaunal). These include a large number of species like some decapod and amphipod crustaceans, sea anemones, echiuroids and certain fish that construct permanent burrow structures in the sea bed. The passage of trawls over muddy grounds collapses and in-fills such systems, creating the need for inhabitants to dig themselves out, perhaps repeatedly on heavily fished grounds, so incurring unknown energetic costs. Such organisms might themselves be the target of the fishery, like Norway lobsters *(Nephrops norvegicus)* in European waters. Bioturbation mounds created at burrow entrances by organisms like thalassinidean mud shrimps or echiuroid worms also become flattened by towing gear across them. Demersal towed gears may deplete shallow-burrowing bioturbators, but the physical act of fishing may substitute for biological activity as the agency of sediment

4

overturn.

Sponges may also live on soft grounds, like the velvet sponge *Hippospongia gossypina* on mangrove peat in the Caribbean Sea. Genera like *Thenea, Trichostema* and *Ciocalypta* have almost disappeared from muddy bottoms in the Mediterranean. A century or more ago, the Mediterranean Sea was the sole commercial source of sponges, but from 1900 onwards sponge fisheries have developed in the Caribbean Sea and Gulf of Mexico, notably for the velvet sponge. It seems that velvet sponges thrive best on detrital food from decaying seagrass leaves, so any reduction in seagrass bed integrity might also feed-back on adjacent sponge populations.

Soft muds under natural conditions are usually characterized by an easily resuspended surface floc of high water content derived both from settlement of organic detritus from the overlying water column and from accumulated faeces of infauna (and pseudofaeces in the case of bivalves). Such flocs may contain specific nutrients (possibly pollutants too) and/or larval settlement factors. Mechanical disturbance could lead to imbalances about which we presently know little. American work has shown that scraping off the top mud layer removes food and inhibits settlement of benthic larvae, and thereby recolonization, in estuarine ecosystems. The resuspension of superficial fine sediments during the passage of trawls over muddy fishing grounds will increase the turbidity of the water which may create problems both for suspension feeders and sight-orientating predators. It may lower the concentration of dissolved oxygen in the water because it enhances provision of surfaces for microbial growth.

4.2 Sand
Sandy sediments can be more consolidated than muds and have greater load-bearing properties. Since they generally occur in shallower, more energetic waters inshore, sands are inherently less stable than muds and the opportunities for organisms to create permanent homes in sand are much less. The infauna tends to be more mobile (bivalves, sea urchins, errant polychaetes), and a diverse array of motile surface-dwellers (brittlestars, whelks, scallops, crabs, hermit crabs) is typical of sandy muds, muddy sands and sands. Though protected by internal or external skeletons, the brittle carapaces, shells or tests of such animals are easily broken by contact with heavy towed gear, and such organisms may be especially vulnerable to trawling. The buried infauna of sandy grounds, mostly small polychaete worms, amphipods and bivalves, are probably least affected by mechanical disturbance, being adapted to living in a shifting substratum.

That said, certain colony-forming organisms, like the honeycomb worm *(Sabellaria spinulosa)*, by virtue of their binding sand grains together to form juxtaposed tubes to live in, can substantially consolidate and roughen off-shore sandy sediments. Anecdotal evidence has suggested that reefs of this worm in the Wadden Sea were deliberately ground-up by fishers using heavy gear to clear the area of net-abrading material prior to deployment of light shrimp trawls.

4.3 Gravel and mixed grounds
Gravel grounds are typically found in areas of wave or current action. Such areas in European waters, although sometimes sustaining relatively low diversity ecosystems, may support dense populations of suspension-feeding fauna including bivalves *(Modiolus modiolus, Limaria hians)* and brittlestars *(Ophiothrix fragilis, Ophiocomina nigra)*. Such animals are easily torn or dislodged from their substratum. For

4

instance, a substantial and well documented *Modiolus* ground to the south of the Isle of Man, Irish Sea, has been fragmented, presumably by intense scallop-dredging activity during the last 20-30 years over increasingly marginal grounds.

Mixed bottoms, such as stony or shelly patches with rocky outcrops between softer patches of sandy or muddy sediments, have traditionally been regarded by fishers as potentially lucrative, but hazardous to gear. However, as productive grounds become overfished, more and more effort tends to be directed to such areas. This has led to the development of gears designed to cope with a rough sea bed, hence the chain mat on beam trawls and 'rock-hopper' modifications to otter trawls.

Gravel and mixed grounds are relatively stable, permitting colonization by slow-growing sessile fauna *(Modiolus, Limaria)* which further bind the shingle. Such organisms themselves become habitats for many other smaller organisms, which provide food for small and juvenile fishes. Colonial organisms like soft corals *(Alcyonium digitatum),* hydroids *(Nemertesia antennina)* or sea fans *(Eunicella verrucosa),* which require solid surfaces for attachment, can never reattach once dislodged. Removal of such species will also carry consequences for a wide diversity of sedentary organisms reliant on such long-lived species to provide a suitable substratum for larval settlement, including commercial bivalve shellfish. Concerns have been expressed recently about the possibility that substantial degradation of the vulnerable shell veneer overlying muds has occurred in scallop-dredged areas of Strangford Lough in Northern Ireland to the detriment of the diverse associated fauna. Changes in the topography of the bottom, however, even on a small scale (e.g. in terms of stone- or shell-over-turn) can lead to reduction in the numbers of refuges available for small animals like squat lob-

sters *(Galathea)* and amphipods to hide from predators, and the reduction of solid surfaces for larval settlement. On the other hand, new surfaces may be made available for colonization.

Demersal trawls covering mixed grounds in the North Sea are commonly contaminated with certain bryozoans *(Alcyonidium).* Their repeated handling can provoke severe allergic dermatitis in fishers, characterized by a painful rash and large weeping blisters.

4.4 Maerl

Maerl is an unusual and fragile bottom type composed of caltrop-like or globose growth forms of calcareous rhodophytes (red algae). Two maerl-forming species, *Phymatolithon calcareum* and *Lithothamnion corallioides,* are listed as needing management protection in European seas under the 1992 EC Habitats Directive (Annex V(b)).

Being living algae, maerl-forming species require access to light for photosynthesis, and thrive best in relatively clear waters. As such, maerl grounds are usually found in shallow-water sites with a high degree of water movement, areas that also support high densities of scallops. In redistributing sediments, scallop dredging can bury and/or smother and smash maerl, modifying the water permeability by clogging pore spaces and impairing oxygenation, and so reducing the scope for the expression of an otherwise highly diverse and characteristic fauna.

Maerl-forming species grow exceptionally slowly, with thalli maybe taking centuries to grow a few centimetres. As a result, maerl beds may take thousands of years to accumulate. One pass with a scallop dredge can create damage that could take decades to reinstate. Fragile infaunal species, like irregular sea urchins *(Spatangus),* and even some large bivalve molluscs *(Venus,*

Maerl is formed by calcareous red algae

Maerl is particularly sensitive to the direct effects of heavy towed fishing gears

4

Figure 9 Coral reefs are complex habitats and easily damaged by the direct impacts of fishing gears. (Photograph: © S.Jennings)

Dosinia) may be damaged in the process as well.

4.5 Coral, coralligène and rocky reef habitats

Hard ground may either be broken or solid rock, or it can be created by living organisms. Coral reefs are the classic example of the latter category (Figure 9). These occur not only in tropical seas (reef-building or hermatypic species), but also penetrate into cold-temperate environments (non reef-building species). Coralligène bottoms occur in the Mediterranean Sea and are complex concretions formed by association of a variety of calcareous red algae, either on rock (coralgal reefs) or on gravel (coralgal banks).

Some fishing methods, such as explosives and drive-netting (see Section 2.6), destroy the physical integrity of hard sea beds, such as coral reefs. Nevertheless, such fishing practices are widespread in Southeast Asia and Micronesia, and actively growing coral reefs, with complex habitat structure and diverse faunas, have often been reduced to rubble in intensively fished areas. Since the abundance and diversity of fishes on reefs is correlated with habitat

Coral reefs grow slowly and recover slowly from damage caused by fishing gears

complexity, damaged reefs rarely harbour a productive or varied fish fauna. Larger species, favoured by fishers, often vacate areas of low habitat diversity, leaving less desirable species, like small wrasse, to proliferate.

The direct impacts of fishing on coral reefs depend upon the intensity of fishing and the stability of the environment. On those reefs situated in areas subject to hurricane damage, the shallow-water coral community is often dominated by faster-growing, opportunistic species and the long-term effects of fishing are often less noticeable. On reefs outside hurricane areas, however, many older corals may be present, and chronic damage from fishing activities can lead to long-term changes in reef structure.

Coral reefs may take a long time to recover from direct fishing effects. While the marks left by a trawl on a sandy sea bed in a shallow tidal sea may be gone within hours, an area damaged by dynamite fishing may remain visible for years. Recovery of reefs relies on the regrowth of damaged corals or coral fragments,

4

and the settlement of coral planulae from the plankton. Most corals only grow a few centimetres each year and even a coral community dominated by fast-growing, branching species would take five to ten years to develop once fishing stopped.

Hand-harvesting of rocky shore plants and animals for human food, bait or whatever purpose, affects the rest of the ecosystem. Harvesting of space-occupying species, like stalked barnacles *(Pollicipes)*, acorn barnacles, mussels or large tunicates (such as *Pyura* taken for bait in South Africa) will open free space (perhaps for decades) and initiate man-made successional sequences. Many invertebrates, once fully grown, gain a respite from marine predation pressures once they become large in size. Selection by humans of large individuals (e.g. limpets, mussels) will run counter to this tendency and also remove the most fecund individuals, lowering reproductive output in species which generally have erratic recruitment anyway, and perhaps tipping any ecological balance for or against other organisms.

On the Canary islands, where limpets are eaten by people and birds, the extinction of the endemic oystercatcher *(Haematopus meadewaldoi)* has been blamed on human overexploitation of limpets. However, exploitation of limpets in Chile inadvertently yielded another resource for humans in the shape of the commercially important alga *Iridaea laminarioides,* which proliferated as a result of reduced grazing pressure. Now limpets are harvested on a 4-year cycle to allow maximum yields of both organisms.

Heavy towed gears can even impact solid rock bottoms detrimentally, by breaking off rock ledges and displacing and overturning boulders. As fishing extends into ever deeper waters, new vulnerable habitats will require considera-

tion. For instance, reefs of the scleractinian coral *Lophelia pertusa* have been found recently off the continental shelf edge of Norway and the Western Approaches. Large (1.5m high) colonies of this species may be 200-400 years old. Reef-like structures constructed by large serpulid tubeworms *(Serpula vermicularis)* and bryozoans are also particularly fragile. While towing gear through such reefs could be regarded as having some positive benefits, through redistribution of fragments enhancing colonization of the surrounding sea floor, few scientists would advocate it.

There have been few studies on the impacts of trawling on hard bottom communities. Work currently in progress off Georgia and South Carolina, USA, adjacent to a national marine sanctuary, suggests that the density of barrel sponges *(Cliona spp.)* is significantly decreased by trawling activities, but damage to other large sponges, octocorals and hard corals is not statistically significant.

4.6 Seagrass meadows

Seagrass habitats are very productive and seagrass meadows help to stabilize sediments in inshore waters. Trawling reduces seagrass cover and plant population density and modifies trophic structure. This has been shown both by studies on *Posidonia* meadows in the Mediterranean Sea, where trawling targets fish and shrimps, and on *Zostera* meadows in the USA where the target species is the bay scallop *(Aequipecten irradians).* Damage is greatest when the towed dredge bag is full, and is exacerbated by propeller scour. Dredges and trawls shear-off leaves, expose rhizomes and dig shoots from the substratum (Figure 10). As a result, the seagrasses may no longer bind sediment effectively, and it begins to erode, leading to habitat loss and increased turbidity. Shoot loss may cause long-term habitat degradation

Seagrass meadows stabilize sediments

Seagrass meadows are damaged by trawls and dredges and take many years to recover

4

(a)

(b)

Figure 10 The contrasting appearance of trawling-impacted (a) and pristine (b) seagrass *(Posidonia)* beds off the Spanish Mediterranean coast.(photographs: © A.A.Ramos-Esplá)

(1-2 years), but leaf removal creates an immediate reduction in scallop larval settlement potential, since post-larval scallops require them for attachment surfaces. Natural recovery rates of shoots are slow even in areas with adjacent seed sources and viable grass beds. As revealed by Australian studies on the effects of seismic testing, the recovery from damage caused by removal of plants (including rhizomes) in mature seagrass communities of *Posidonia* and *Amphibolis* is very slow, taking perhaps 60-100 years.

However, it should not be overlooked that natural forces too can wreak havoc in seagrass beds. Some years ago, it was reported from Chesa-

4

peake Bay that cownose rays *(Rhinoptera bonasus)* were responsible for widespread habitat destruction to eelgrass beds *(Zostera marina)*, uprooting the plants wholesale to gain access to their food (infaunal clams, *Mya arenaria)*.

Seagrass habitats are of considerable importance as a food source for certain threatened animals, for example the green turtle *(Chelonia mydas* hence the 'turtle grass' *Thalassia)*, the West Indian manatee *(Trichechus manatus)*, and the dugong, *(Dugong dugon)*, so their maintenance is of high conservation significance. They also serve as nursery grounds for many commercial species, such as the bay scallop in the USA (above), young of the year northern Atlantic cod in Newfoundland and many commercially-fished wrasse (Labridae) and sea bream (Sparidae) species in the Mediterranean Sea.

4.7 Kelp forests

Kelp forests provide important habitats for some commercial shellfish, including sea urchins, abalones, lobsters and crabs that are exploited either using fixed gears or by divers. They may also represent important nursery grounds for commercial fishes. The extinction in the 1700s of Steller's sea cow *(Hydrodamilis gigas)* through overhunting represents the irreplaceable loss of a giant herbivore whose impacts on the character of the kelp forest will never be known.

Kelp beds are usually not reduced by direct fishing activities, although kelp itself is harvested extensively in some parts of the world for alginate extraction. However, some kelp forests, like those formed by the (now rare) *Laminaria rodriguezii* in deep waters (60-150m), have almost disappeared from the Mediterranean Sea due to intensive trawling. Forests of this species once occurred commonly in the Mediterranean Sea, but they are now restricted to non fishable areas around the Balearic islands. For target

species with demersal or adhesive eggs (e.g. octopods, some elasmobranchs), fishing or kelp-harvesting gear may damage or remove deposited egg capsules.

Kelp beds produce particulate organic detritus that benefits adjacent, and not so adjacent ecosystems, including the deep sea. Changes in the balance between producers, consumers and their macroparasites in this ecosystem can lead to dramatic changes in habitat. Thus, a population 'explosion' of sea urchins *(Centrostephanus rodgersii)* in New South Wales, Australia may have resulted from the overexploitation of their only predators, the grey nurse shark *(Carcharias arenarius)* and the blue grouper *(Acheoerodus viridis)*. This led to overgrazing of benthic algae and associated rock-attached flora and fauna resulting in the generation of barren grounds.

Heavy sedimentation, caused by bottom disturbance from dredging activities, settling onto photosynthetic fronds may further impair kelp *(Laminaria)* growth beyond the indirect effect of light reduction caused by increased turbidity. As with coral reefs and seagrass meadows, the integrity of kelp beds is also important geomorphologically in moderating the force with which storms batter coastlines.

4.8 Sea mounts

Fish often aggregate over seamounts, perhaps because diurnally migrating plankton, entrapped as they are advected over the seamount, attract consumers. Seamounts may also be places for spawning aggregations of fish. The discovery of such deep-sea resources in recent years has prompted a modern 'gold-rush' of exploitation, with work on its impacts lagging far behind.

Recent work off the Chatham Rise, New Zealand, in waters between 662 and 1524m that

Sea mounts can be productive fishing grounds

But many of the slow growing species found there are very vunerable to fishing

25

4

have been fished commercially since 1978, compared benthic invertebrate by-catch derived from the fishery for orange roughy *(Hoplosthetus atlanticus)* from flat areas and hills (small sea mounts). By-catches differed significantly between these areas. Dominant groups from the flats were sea cucumbers, starfish and shrimps, whereas groups most commonly recorded from hills were corals. The largest by-catch volume comprised corals from hills: Scleractinia *(Goniocorella dumosa)*, Stylasteridae *(Errina chathamensis)* and Antipatharia *(Bathyplates platycaulus)*. The suggestion is that such deep-sea corals might take >100yr to recover from trawl damage.

Relatively pristine seamounts newly charted in the Australian Fishing Zone off southern Tasmania were accorded temporary protection in 1995, with research ongoing to determine the level of permanent protection, if any, that they should be afforded.

Deep-water trawl fisheries, in particular for orange roughy, impact detrimentally on other deep-water fish too, like smoothheads (Alepocephalidae) in waters <1,200 m, and shark populations at all depths. The flesh of smoothheads has a very high water content, so these `nuisance fish' have no commercial value; yet researchers have found they constitute up to 61% of trawl catches from the continental slope West of Scotland in 1,000-1,200 m of water. They do not survive when discarded.

In general, deep-water species are more delicate than their shallow-water counterparts and many fish also die through scale loss, even if they escape through the meshes of trawl nets. In a study of the discards of vessels fishing in 350-1,300m in the Rockall Trough (North Atlantic), the ratio of landings to discards for round-nosed grenadier *(Coryphaenoides rupestris)* was approximately 1:1.

Life histories and behaviour determine vulnerability to fishing

4.9 Relative vulnerability

The organisms most at risk from towed fishing gears (besides the target species) are long-lived, slow-growing species: either surface-attached organisms like sea fans *(Swiftia, Eunicella,* axinellid sponges), horse mussels *(Modiolus),* Ross coral (actually a bryozoan, *Pentapora foliacea)* or sedentary infauna, like the shallow-burrowing bivalves *Arctica islandica, Astarte* spp. and *Laevicardium oblongum* (Table 2). Deep-burrowing species, like certain bivalves *(Lutraria),* polychaetes *(Sabella)* and sea anemones *(Cerianthus)* may be able to retract themselves quickly enough when sensing the approach of towed gear, so avoiding damage. Shallower burrowers *(Venus, Ensis)* and tube-dwellers *(Chaetopterus)* run the risk of being sliced off *in situ* as they span the depth range of penetration of the gear.

The Southern Ocean has a rich benthic fauna with a large number of long-lived, slow-growing forms. The extent of impacts by towed demersal gears in this area, like the destruction of habitats and fish spawning grounds, is unknown. However, here again effects are likely to be long-lasting owing to the fragility and slow recovery rate of the biota at prevailing low temperatures. In order to minimize the impact of trawling on non-target species and on the sea bed, bottom trawls have been prohibited for mackerel icefish around South Georgia.

The behavioural reactions of species will contribute to their relative vulnerability to fishing. The burrow-emergence behaviour, and hence catchability, of Norway lobsters is light-dependent. To maximize catches, trawls are fished at different times of day in different depths of water, since appropriate light conditions for emergence occur variously at night (in shallow water), at dawn and dusk (intermediate depths) or during daylight hours (in deep water). Consequently, non-target organisms

4

Table 2 Biological characteristics affecting the vulnerability of species to human exploitation (adapted from Dye *et al.*, 1994).

	Vulnerability	
Characteristics	Increasing	Decreasing
a) Turnover		
Longevity	Long	Short
Growth rate	Slow	Fast
Natural mortality rate	Low	High
Production: Biomass	Low	High
b) Reproduction		
Reproductive effort	Low	High
Reproductive frequency	Semelparity	Iteroparity
Age at sexual maturity	Old	Young
Sexual dimorphism	Yes, if female larger	No (or yes, if male larger)
Fertilization	Internal	External
Sex change	Occurs (esp. protandry)	Does not occur
c) Capacity for recovery		
Regeneration from fragments	Does not occur	Occurs
Dispersal	Short-distance	Long-distance
Competitive ability	Poor	Good
Colonizing ability	Poor	Good
Recruitment by larval settlement	Irregular, low level	Predictable, frequent, intense
d) Ease of exploitation		
Dispersion	Aggregated	Dispersed
Niche breadth	Narrow	Wide
Habitat	Accessible	Inaccessible

with pronounced diel behavioural cycles will also be impacted differently by different types of trawls at different times of day, depending on ground type and target species.

Many deep-sea organisms are slow growing and long lived, with low recruitment rates. These properties mean that deep-ocean ecosystems are especially vulnerable to the emergent threats of unsustainable 'gold rush'-style exploitation of exotic fish resources. This needs to be recognized urgently, since both direct and indirect damage caused by asset-stripping such resources may be both catastrophic and irreversible.

In shallow waters, sensitive habitats like coral reefs, maerl beds, seagrass meadows and some mixed grounds, which support high biodiversities function as important nursery grounds to a range of commercial and non commercial species. They need to be respected by fishers, and protected from harm. This may be achieved by the creation of appropriately sited marine reserves (Section 7.1) or by regulation restricting exploitation of such vulnerable habitats to the use of specified set gears. As explained above, other shelf sea-bed types are more resilient to disturbance. In particular, shallow sandy and muddy grounds generally support organisms adapted to living on and in a dynamic substratum (Figure 11). What we consider as an acceptable level of damage depends on how we balance the benefits of fish protein against the changes that fishing causes in the marine ecosystem.

4

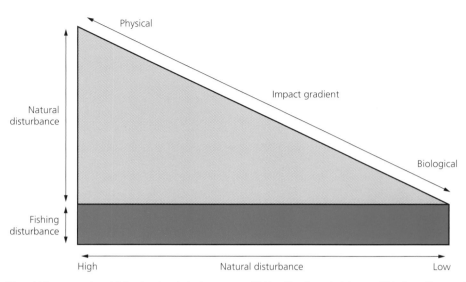

Figure 11 A conceptual model showing the relative importance of fishing disturbance in habatats which also suffer natural disturbance. As the intensity of natural disturbance declines, fishing accounts for a greater proportion of total disturbance. (redrawn from Jennings & Kaiser, 1998).

Summary

Most intensive fishing occurs in highly productive shallow shelf areas. The direct impacts of fishing gears are easier to detect on stable mud sea beds than on mobile sand or gravel. Complex slow-growing biogenic habitats such as maerl beds, coral reefs and seagrass beds are particularly vulnerable to gear damage. The relative effects of fishing depend on the intensity of fishing and the existing levels of natural disturbance in the marine environment.

5 Effects on non-target organisms

5.1 Sea birds

The large quantities of fish and offal discarded in the northeast Atlantic support artificially high populations of surface-feeding seabirds, notably fulmars, kittiwakes *(Rissa tridactyla),* great black-backed *(Larus marinus)* and herring gulls *(L. argentatus).* Discards and offal represent 30% of the total food consumed by seabirds in the North Sea. The total number of scavenging seabirds in the northeast Atlantic has increased from about 37,000 breeding pairs in 1900 to about 614,000 pairs today. European populations of fulmars have expanded considerably over the last century, hand-in-hand with the historical expansion of the whaling (now curtailed) and fishing industries (Figure 12).

Discarded by-catches support seabird populations

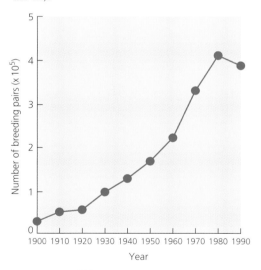

Figure 12 Pairs of fulmars breeding on the east and northeast coasts of Britain since 1900; their increase has coincided with increasing rates of discarding in North Sea fisheries (data from Furness, 1992).

Seabirds actively seek out fishing vessels as an easy source of food, travelling from distances in excess of 10km away, depending on species. Consequently, they are able to obtain a relatively constant supply of food from trawlers.

There is a clear dominance hierarchy among seabirds feeding at fishing boats in UK waters. Fulmars displace all other species and tend to select offal. Gannets *(Sula bassana)* and great skuas *(Catharacta skua)* dominate gulls. Great black-backed gulls dominate herring gulls, and kittiwakes are only able to feed on scraps that other species ignore. Seabirds that followed Dutch shrimp trawlers were reported to consume 41 % of flatfish, 79 % of roundfish and 23 % of four invertebrate species discarded experimentally. It was calculated that North Sea discards from the shrimp trawler fleet of Lower Saxony alone were sufficient to meet the energy demand of 60,000 seabirds for the whole year.

Reducing the supply of discards to seabirds, and thereby forcing these large birds to switch diet, may have a severe impact on other seabird populations. For instance, great skuas and great black-backed gulls are likely to switch their predatory attentions to puffins *(Fratercula arctica),* kittiwakes and storm petrels *(Hydrobates and Oceanodroma* spp.). Severe reductions in discarding can reduce the reproductive success of seabirds; effects on breeding success or population size of scavenging seabirds have been well documented for several gull species in the Mediterranean Sea. For instance, the largest

5

colony (70% of world population) of the threatened Audouin's gull *(Larus audouinii)* occurs in the Ebro delta (northwest Mediterranean). These gulls scavenge trawler discards. After seasonal closure of the fishery for two months in 1993, the breeding success of this gull decreased by 48%.

Direct detrimental effects of fishing on seabirds, mainly on underwater swimming and diving species, are also evident. Lost and discarded nets remain a threat (Section 3.1), although this threat is likely to be small compared with losses attributed to entanglement in active gear. For instance it is estimated that between 214, 500 and 763, 000 seabirds are killed annually by Japanese drift nets in the North Pacific Ocean. Seabird by-catch is highest in the vicinity of major breeding colonies and decreases rapidly with distance from such sites.

In some circumstances net-mortality may constitute the greatest source of mortality in local populations. For instance, about 1,600 razorbills *(Alca torda)* breed in Newfoundland. On average, 12.4% of this population was killed annually in nets between 1981-84. This additional mortality rate exceeded most estimates of annual adult mortality of razorbills in stable populations in the North Atlantic (estimated at about 10%) and such results gave rise to concerns about the viability of this population. In the same region, as much as 16% of some local populations of adult common guillemots *(Uria aalge)* were killed in gill nets in 1982. A substantial proportion of seabirds killed (perhaps 5-13%) float free of the nets upon their retrieval and failure to take this factor into account can lead to considerable underestimations of mortality.

Even small incidental catches of sea birds in net fisheries can sometimes have disproportionate conservation significance. For example, among

bird species known to be caught incidentally in offshore fishing nets in the Pacific northwest, only the elusive marbled murrelet *(Brachyramphus marmoratus)* is listed under the US Endangered Species Act as well as the Migratory Bird Treaty Act. The 200 or so marbled murrelets killed each year in Berkeley Sound, British Columbia represents a considerable take from the population resident in the area.

Many abundant seabirds are rarely caught in nets because they are surface feeders with limited or no diving ability (e.g. storm petrels, *Larus* gulls and kittiwakes). Plunge-diving gannets, however, have been caught in salmon gill-nets off Newfoundland, and substantial numbers have been entangled in ghost fishing nets off Helgoland.

Little is known about the age-classes of seabirds caught in fishing nets. Both adult and subadult tufted puffins *(Lunda cirrhata)* have been caught, but the proportion of adults to subadults varied greatly with the location of the sample. Mesh size is an important consideration. Relatively few birds were captured in nets with mesh sizes <82mm, and more were caught in nets with meshes between 106 and 132mm. Unfortunately, this represents the mesh size used by Japanese boats in the North Pacific salmon gillnet fishery. One of the reasons why many short-tailed shearwaters *(Puffinus tenuirostris)* become entangled in gillnets may lie in the similarity of its feeding habits and distribution to those of Pacific salmon, both of which prey on euphausiids near the surface.

The southern bluefin tuna *(Thunnus maccoyii)* and Patagonian toothfish longline fisheries have had dramatic effects on mortality of wandering albatrosses *(Diomedea exulans)* and other albatross species. Estimated death rates range from about 5 birds per 10,000 hooks. Prior to the introduction of mitigation mea-

Seabirds may be trapped and killed in fishing nets

5

sures in 1989 up to 44,000 albatrosses were estimated to have been killed by the global Southern Ocean bluefin tuna fleet which deployed some 107 million hooks annually. Since then the number of hooks has been reduced by fisheries regulation and measures to deter seabirds introduced (Figure 13). In 1990 off Tasmania, Japanese tuna longliners reduced seabird by-catch by 88% (cf. 1989) by using seabird deterrent devices. In 1995, longlining was listed as a threat to populations of wandering albatross under the Australian Endangered Species Protection Act 1992. Modelling studies have suggested that unless the longline fishery in the southern oceans is reduced to below 41 million hooks set per year, or effective mitigation measures adopted, then the wandering albatross will decline to extinction. It has been estimated that, in the 1980s, the southern bluefin tuna fishery lost about £2 million per year in potential fish catches as a result of bait loss to seabirds.

Although the demersal longline fishery for Patagonian toothfish uses mitigation measures to reduce seabird by-catch in CCAMLR waters, the much larger illegal and/or unregulated fishery on this species was estimated to kill up to 140,000 seabirds (mainly albatrosses

and white-chinned petrels) in 1997. Although there are few data, it seems likely that the Norwegian, Icelandic and Alaskan demersal longline fisheries catch even more seabirds than the tuna fisheries but most of the birds caught are fulmars, and no impact on fulmar population size has yet been detected.

Given the small world populations of Amsterdam albatross, *Diomedea amsterdamensis* (<100 birds; IUCN status 'Critically Endangered'), and short-tailed albatross, *D. albatrus* (<1,000 birds; status 'Vulnerable'), the impact of removing even a few mature individuals from world breeding populations of 13 and 158 pairs, respectively, would be considerable. Satellite tracking shows that the former species forages over known tuna and toothfish fishing grounds. The latter has already been caught in small numbers by longliners operating in Hawaiian waters (see also 7.2).

Since fishers and seabirds may compete for the same resource, it is often assumed that large industrial fisheries will deprive seabirds of a food supply (Figure 14). However, empirical evidence demonstrates, counterintuitively, that industrial fishing does not consistently cause seabird population declines. In the North Sea,

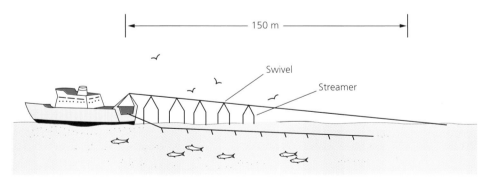

Figure 13 Streamer line designed by CCAMLR to discourage birds from scavenging baits during deployment of longlines by vessels setting at 5 -10 knots.

Fisheries compete with seabirds
for prey fish

Fishers compete with wading
seabirds for shellfish

Marine mammal by-catches have
led to long-term decreases in
their abundance

for instance, the industrial fishery for sandeels *(Ammodytes* spp.) now takes around 700,000 to one million tonnes of sandeels each year. Although sandeels are the main food of most seabirds in the North Sea in summer (and monitoring seabirds there is more intensive than anywhere else in the world), there have not been widespread effects on seabird numbers or breeding success that can be attributed to the industrial fishery. However, there is increasing evidence that industrial fishing in some areas leads to localised depletion of sandeels and this may have consequences for bird populations feeding in these areas.

Overexploitation of benthic epifaunal shellfish stocks, like scallops and mussels, may adversely impact populations of diving ducks, like eider *(Somateria mollissima)*, scoter *(Melanitta* spp.*)* and long-tailed duck *(Clangula hyemalis)*. Eiders routinely forage offshore in water 20-30 m deep and, in Northern Norway, have been shown (in association with king eiders) to be capable themselves of removing the entire annual production of prey bivalves *(Chlamys islandica)* in an area. In years with a low stock and high market price it is profitable for the Dutch cockle *(Cerastoderma edule)* fishery to take undersized animals, including those as small as 16-18 mm long. Since the minimum energetically profitable size for oystercatchers is around 15 mm, the birds cannot really escape the effects of fishing by taking smaller cockles. Because the best Dutch cockle beds are nearly completely fished out and cockle condition is poorer in marginal areas, the cockle fishery is easily able to reduce by halve the food supply of oystercatchers in this major wintering area.

5.2 Sea mammals

The annual death toll of hundreds of thousands of dolphins, mainly pantropical spotted dolphins *(Stenella attenuata)* and spinner dolphins *(S. longirostris)*, in the Eastern Tropical Pacific tuna purse seine fishery during the 1960s and

Figure 14 Seabirds such as these blue-footed boobies may feed on fishes that are also caught by fishers (Photograph:S. Jennings).

5

early 1970s is the most dramatic example of the by-catch problem. By 1980, eastern spinner dolphin population sizes were reduced to around 20% of pre-exploitation levels and spotted dolphin populations to 40-50% of pre-exploitation levels (Figure 15). Arguments about how much to reduce dolphin mortality in this tuna fishery became known as the "tuna-dolphin" debate. Subsequent changes in fishing techniques (Section 7.2) have now dramatically reduced mortality of dolphins, but unintentionally they have exacerbated the by-catch of sharks, juvenile billfish (sailfish and marlins; *Istiophorus, Tetrapturus* and *Makaira spp.),* sea turtles and other non-target species. In 1993 and 1994, the total by-catch for this fishery was estimated at almost 6.6 million animals, of which only a few thousand were dolphins. Regulations imposed as a result of the US Marine Mammal Protection Act (1972) have included specification of particular gear modifications and fishing practices (to facilitate the so-called backdown procedure to release encircled dolphins), banning net sets at sundown because of their higher dolphin mortality rates, instituting an observer programme to monitor dolphin

Marine mammals compete with fishers for prey

mortality, and placing an embargo on tuna imports from countries not complying with these requirements. Although that embargo was challenged successfully by Mexico in 1991 under the GATT agreement, the result was the International Dolphin Conservation Programme set up in 1992. This Programme, co-ordinated by the Inter-American Tropical Tuna Commission (IATTC), provided for 100% observer coverage of the entire international tuna purse seine fleet and set mortality quotas for individual vessels. These measures reduced mortality of dolphins from 100,000 in 1986 to 4,000 in 1993.

Human activities have caused changes in the behaviour of marine mammals. Dolphins exploited in the Solomon Islands avoid vessels more than the same species in other unexploited populations. Indeed, swimming away from vessels may constitute a significant proportion of the daily energy consumption of animals during the fishing season. Such responses are understandable when it is realized that individual dolphins may be captured several times daily by tuna purse seiners resulting in physiological stress. On the other hand, some whales and dolphins follow ships. Steller's sea lions *(Eumetopias jubatus)* have been observed following factory trawlers and consuming discards and processing wastes released by these vessels. Seals, together with sharks and diving seabirds, might also conceivably profit from consuming fisheries discards as they fall through the water column; but the main interactions between marine mammals (whales, porpoises, dolphins, seals etc) and fishing gears are likely to be detrimental.

Entanglement among net debris is especially significant. The 13% incidence of dead fur

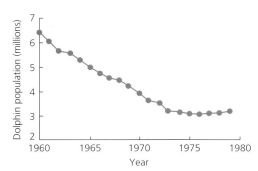

Figure 15 Abundance of Pacific spotted dolphins from 1960 to 1979 (data from Allen 1985).

Rare turtles are threatened by
shrimp trawls

seals *(Callorhinus ursinus)* reported in nets lost off Japan and the estimate of 4,000 net fragments within 370 km of Japan's northern coast lead to an estimated 533 fur seals entangled and drowned in nets lost in this area alone. In 1990, the IWC considered the problem of cetacean mortality in passive fishing gear to be sufficiently important as to warrant sponsoring a Workshop on the topic. At the resultant 1991 Workshop scientists noted that seven out of 54 species population regions (SPRs) had mortality rates for passive gears that were determined to be "not sustainable".

Marine mammals are often viewed unsympathetically by fishers, since they compete with them for target species, damage and steal from fishing gear and scare fishes, dispersing schools and stopping target fish feeding. Fur seals and sea lions are often captured in nets while trying to take captured fish from nets. The fact that the number of Cape fur seals *(Arctocephalus pusillus pusillus)* trapped increased with age in South African waters suggested that scavenging from fishing vessels might be a learned behaviour. Also, the sex ratio of those captured was significantly different (male biased) compared with the population as a whole. In the Baltic Sea, most seals (predominantly Atlantic greys, *Halichoerus grypus)* snagged in fishing gear were caught in eel traps and turbot nets, and pups of the year made up about half the total by-catch. Endangered Hawaiian monk seals *(Monachus schauinslandi)* have also been reported entangled in fishing nets. The vaquita *(Phocoena sinus)*, a species of harbour porpoise inhabiting the Gulf of California, and considered to be the most endangered marine cetacean species in the world is taken as a by-catch in the gill net fisheries in the Gulf. High mortalities caused by capture in passive gears affect other cetacean populations, including the baiji, or Chinese river dolphin *(Lipotes vexillifer)*, the Indo-Pacif-

ic hump-backed dolphin *(Sousa chinensis)*, two populations of bottlenose dolphins *(Tursiops truncatus)* off the coast of South Africa, harbour porpoise throughout the Northwest Atlantic, and striped dolphin *(Stenella coeruleoalba)* in the Mediterranean Sea.

Mortality of northern fur seals *(Callorhinus ursinus)* from net entanglement was thought to contribute significantly to their population decline in the 1970s and early 1980s. In 1989, similar entanglement rates for Antarctic fur seals *(Arctocephalus gazella)* resulted in 5,000-10,000 presumed mortalities at South Georgia. This led to a major publicity campaign aimed at Southern Ocean fishers and due to this, and the CCAMLR ban on the use of polypropylene packaging bands, entanglement rates dropped to 50%, and latterly to 20%, of those previously recorded, though fishing effort also decreased concurrently.

The British and Irish gill net fishery in the Celtic Sea has recently been estimated to catch 2,200 harbour porpoises annually. This estimate excludes by-catch from French boats and from smaller British and Irish boats, which also contribute substantial fishing effort. The catch of 22,000 porpoises equates to an average rate of 7.7 porpoises per 10,000 km hours of net immersion, representing 6% of the estimated population of the area. This greatly exceeds the International Whaling Commission's (IWC) threshold for concern for sustainability of by-catch which is an annual loss rate of 1% of population size, and the ASCOBANS criterion that annual bycatch above 2% of population size is unacceptable.

5.3 Sea turtles

Shrimp trawl by-catch has been identified as the most significant source of sea turtle mortality in the USA. Since all five species of sea tur-

5

tles found in US waters are listed either as threatened or endangered, effective controls on all additional mortality causes are required. The by-catch toll has put the Kemp's ridley turtle on the endangered species list. The abundance of stranded sea turtles and the intensity of shrimp fishing in the USA varies in both space and time, but in only two out of eight US studies did times and locations of turtle strandings positively correlate with fishing effort. The problem is that such contemporaneous analyses take no account of historical changes in turtle abundance in relation to past shrimping and other mortality factors. Thus the relationship between strandings and fishing effort is more complex than the simplistic argument that more shrimping equals more strandings. In the western Mediterranean Sea, 20,000 endangered turtles (like loggerheads *Caretta caretta* or leatherbacks *Dermochelys coriacea*) are caught inadvertently in the swordfish longline fisheries each year.

Even if turtles can be released from fishing gears, a study showed that, off the coast of Queensland, their survival rate varied with species, size, condition (before capture), degree of injury and exhaustion, depth of capture and trawl duration. It was thought that 15 to 40% of sea turtles inadvertently captured subsequently died. In 1987, it was established that there was a highly significant positive correlation between tow time and incidence of death in entangled sea turtles. Death rates were near zero until tow times exceeded 60 minutes, rising to 50% for tow times of 200 minutes, an unsurprising consequence of underwater entanglement of an air-breathing animal. Death rates never reach 100%, however, since some turtles are captured just before hauling of the gear. A turtle death rate of about 10% has been reported in trawl fisheries off North Carolina, USA.

An additional consideration is the problem posed by repeat captures. Thus, off the coast of Georgia, USA it was common to recapture the same turtle in the same day, often immediately following its release. Such individuals are probably more prone to drowning than unstressed turtles. This is particularly likely to be a problem in places where high trawling intensity occurs over restricted areas.

Juvenile sea turtles drift passively with ocean currents for maybe 3-5 years, during which time they come into contact with buoyant debris. As a result they suffer entanglement, entrapment and, possibly, damage to the alimentary canal after ingestion of fragments. In particular the front flippers and head become easily entangled with monofilament nets, buoy ropes or discarded twine. The marked tendency of leatherback turtles to ingest plastic bags has been attributed to their being misidentified for jellyfish which are one of their staple foods. This species seems to be rather indiscriminate in its feeding. One was found in 1980 off New York with 180m length of heavy monofilament line in its gut.

5.4 Sea snakes

Of the 16 sea snake species that occur in the Gulf of Carpentaria (Australia), ten species have been caught in trawls (mostly prawn trawls). In 1991, the estimated catch of sea snakes in that area was 120,000. Whether this represents a threat to their populations is unknown, however, since so little work has been done on these species. Their mortality rate post-capture was between 10-42%, varying considerably between species, with trawl duration and post-capture handling. Prawn fishermen from that area reported that unidentified species of sharks followed their ships and, although feeding on other 'trash' animals discarded, did not eat the sea snakes (chiefly *Hydrophis elegans*). The established unpalata-

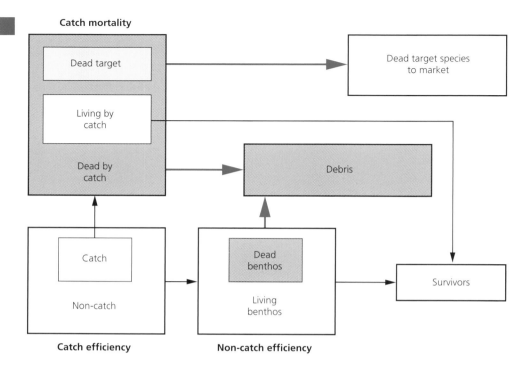

Figure 16 Direct effect of beam trawling on demersal fish and benthic invertebrates as related to (a) the catch efficiency: i.e. the number of fish and invertebrates caught in the nets divided by the total number of animals present before fishing, (b) the catch mortality, i.e. the number of dead fish and invertebrates in the catch divided by the total number of animals in the catch, and (c) the non-catch mortality, i.e. the number of dead fish and invertebrates in the trawl track divided by the total number of animals in the trawl track after fishing. The bold arrows represent the fluxes of dead animals, while the narrow ones indicate the fluxes of (initially) living animals (modified after Lindeboom & de Groot, 1998).

bility of sea snakes to aquatic predators may therefore increase their chance of recovery once returned to the sea.

5.5 Benthic scavengers

We have seen how much of the material discarded by fishers becomes available to scavenging species, according to processes summarised in Figure 16. A suite of, largely facultative, scavengers responds to the presence of carrion falling to the sea bed in a sequence that reflects their different motilities. Demersal fishes move quickly into trawl-disturbed areas to feed on damaged animals, but as yet it is unclear whether pelagic fish consume many of the discards falling from the surface. The succession of scavengers that occurs at carrion will depend on the type of ground, the local abundance of different species, their speed of movement and prevailing current conditions. Foraging will normally be up-current or zig-zagging across current, and the dispersion of the odour plume will vary with tidal state. Different organisms may be active during the day and night, especially in shallow water. Typically such macrofaunal scavengers include starfish, brittlestars,

Scavengers feed on waste discarded during fishing activities

5

hermit crabs, amphipods and isopods.

The extent to which these organisms interact and select their food is a topic of current research. It may be that there are some obligate scavengers in the sea, perhaps focused on crustacean carrion. Certainly the speed with which corpses on the sea bed disappear is testimony to the efficiency of scavenging organisms. A dead fish can be reduced to skin and bone overnight by cirolanid isopods in coastal waters, and a dead dolphin can be skeletonized in the deep sea within a week, mainly by lysianassoid amphipods.

Evidence has been sought to support the hypothesis that the populations of benthic scavengers might have expanded in response to carrion made available from fishing activities, but the evidence is weak in contrast to that for surface-feeding seabirds. In the North Sea, the annual amount of carrion produced by fishing activities accounts for a maximum of 10% of the annual food consumption by scavenger populations. Discarded material settling through the water column will be broadcast sparingly over a very wide area and provoke only local, chance consumption. None of these benthic species can follow their food sources as gulls can.

It is possible though that large numbers of small mobile scavengers (like amphipods and isopods), exploiting the artificially elevated supply of damaged animals on the ground (as in the wake of trawls), might be responsible for enhancing disease transmission between individuals in commercially valuable benthic species. For instance, the transmission pathway for the pathogenic syndinean dinoflagellate *Hematodinium* sp. which infects Norway lobsters in British and Scandinavian waters (at least) has yet to be established, but could well involve such organisms.

Summary
Discarded by-catches and fishery waste provide an increasingly important food source for seabirds, and discards that sink provide food for benthic scavengers. While some birds benefit from discarding others suffer mortality as by-catch. Fishers may compete with birds for fish or shellfish prey. Marine mammal and turtle by-catch mortality has been sufficiently high to reduce the abundance of some species.

6 Community and ecosystem responses

6.1 Community diversity

The marine benthos harbours the greatest phyletic diversity of any ecosystem on Earth (35 phyla of which 11 are endemic). Despite the fact that there are some rare phyla containing only a few species it is extremely unlikely that present environmental impacts, such as fishing, will lead to any significant reduction in phyletic diversity. For example, Spanish researchers have found no differences at phylum or class level between trawled and unfished seagrass meadows, but differences were significant at family and species levels.

The most dramatic effects of fishing on diversity and community structures occur at the outset of exploitation. However, once systems enter a fished state, diversity and production overall are often relatively stable, despite changes in fishing intensity. As a result, studies undertaken on systems where fishing preceded scientific study will suggest (possibly wrongly) that those systems have not been affected markedly by fishing. Long-term and large-scale studies (i.e. longer than a human lifespan) that ideally should include unfished control sites, are the best way to elucidate fishery impacts on community structure and diversity.

There is good evidence of local reduction in species richness following fishing. Comparisons between fished and unfished (marine reserve) areas on coral reefs in many regions of the world have shown consistently that species richness is higher in the unfished areas. On Fijian reefs, for example, a marked decrease in the biomass and diversity of piscivorous fish has been reported with increasing fishing effort. Fishing may also lead to indirect changes in the species diversity of fishes or invertebrates by affecting predator-prey relationships or changing habitat structure (Section 6.4).

However, these effects are not consistent and recruitment remains the primary structuring force in coral reef ecosystems. Some of the reported differences between fish communities on coral reefs and in temperate waters may be due to the different scales at which fisheries scientists collect data rather than to any fundamental differences in the responses of these ecosystems to fishing.

6.2 Habitat structure

Little attention has been paid, to date, to the impact of towed gears on the fundamental physical structure, e.g. granulometry, surface roughness or sediment stratification, of sea beds. Trawls towed over sand stir sand particles into suspension, but these settle relatively quickly. Over mud, however, disturbed silt may take hours to settle during which time it may become dispersed by water currents over a wider area. In deeper waters which are not subject to strong wave and current action, sediment suspended by trawls may be the main source of sediment in the water column (Figure 17). Resettlement of particles, with associated smothering potential, thus takes place over a much larger area than that directly disturbed by the passage of the gear. An otter board towed over soft ground fluidizes it resulting in smaller animals often being displaced laterally through movement of piled-up spoil material in front of the door.

6

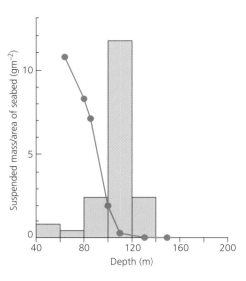

Figure 17 The mass of sediment put into suspension each month by trawlers (bars) on the muddy seabed of the Middle Atlantic Bight from January to March 1985. Circles indicate the mean mass of sediment put into suspension by currents. Below 100m, trawling puts more sediment into suspension (Data from Churchill, 1989).

They may be deflected in this sedimentary 'bow-wave' rather than run over. What may once have been a smooth bottom is then marked by a scour furrow edged with a linear berm feature.

On coral reefs, any large-scale reductions in the structural heterogeneity of the bottom, will lead to substantial impoverishment of fish communities, both in terms of diversity and standing stock. The abundance of reef fishes is correlated with both topographic relief and micro-scale heterogeneity. Some of the most dramatic indirect effects of fishing have been observed on Kenyan and Caribbean coral reefs where fishing has caused the reef ecosystem to shift from a coral-dominated to an algal-, or sea urchin-, dominated phase. Here, herbivorous fishes and sea urchins are important algal grazers and such grazing by fishes clears space for coral settlement and improves the growth and

survival of young coral colonies. Sea urchins are less beneficial to corals since they erode the reef matrix as they feed and prevent further coral settlement. Moreover, by dramatically reducing algal biomass they can arrest the reestablishment of herbivorous fish populations which cannot compete for food with the sea urchins.

Fishers on coral reefs target herbivorous fishes and species such as triggerfishes *(Balistes* spp.) and emperors *(Lethrinus* spp.) which feed on sea urchins. As a result, sea urchins are released from both fish competition and predation, and become the dominant algal grazers, to the detriment of the reef. Once the reef ecosystem enters such a sea urchin-dominated state, the change is unlikely to be reversed. Even in the Caribbean Sea, where disease led to the obliteration of the sea urchin population, herbivorous fishes were so scarce following intensive fishing, that thick mats of algae colonized many of the reefs and prevented coral regrowth.

In temperate ecosystems it has been argued that intensive fishing for at least the last six decades has left the macrobenthic fauna in places like the Dogger Bank (North Sea) adjusted to trawl disturbance. Towed gear is by no means the only physical impact on sea-bed assemblages. Its effects need to be seen in both a physical and biological context: storms, ice scour, grey whale *(Eschrichtius robustus)* and cownose ray feeding (Section 4.6) in different inshore waters and turbidity currents, effectively underwater avalanches, in the deep sea are natural forces that can create immense structural disruption to benthic ecosystems on different spatial and temporal scales. Equally, however, removal of bottom-stabilizing organisms, like the byssate bivalves *Modiolus* or *Limaria*, by fishing activities can destabilize sea-bed sediments making them more susceptible to tidal shear forces.

Water quality may be affected by discarding

Fishing can lead to shifts in habitat structure on coral reefs

6

There are strong ecological links
between benthic and pelagic
communities

practices. A scallop stock inhabiting the Bass Strait, Australia was essentially eliminated within nine months of the start of a commercial fishery. Much of the problem was related to an infection caused by decomposing scallops which had been crushed or damaged incidentally by the scallop trawls. It was estimated that 4-5 times as many scallops were damaged as landed. Ground-poisoning due to oxygen depletion induced by such sources of decomposition would clearly be a cause for concern. This might be a particular problem in seas, like the Baltic Sea, with restricted circulation patterns and poor flushing rates, which may experience anoxic events already exacerbated by eutrophication. Although unconfirmed, an alleged dumping over 60 days of some 47,800 t of discards in the New Zealand west coast hoki *(Macruronus novazealandiae)* fishery was calculated to reduce oxygen saturation to 45-55%. Moreover, during the mid-1970s there were complaints from demersal fishers about dead blistered mackerel polluting the Cornish fishing grounds, after excess catches by pelagic fishers had been dumped (either in a heap by trawlers or scattered after 'slippage' by purse seiners; section 2.4).

Certain Northeast Atlantic Norway lobster sub-populations (stocklets) are composed of dense populations of small animals. When exploited these are usually tailed at sea and the 'heads' thrown overboard. The presence of these discarded heads on the ground has been found to inhibit *Nephrops* movement, so 'spoiling' the ground.

6.3. Benthopelagic coupling

Events in the benthos are coupled to those in the plankton and vice versa. Anthropogenic impacts, including overexploitation, of oysters *(Crassostrea virginica)* in Chesapeake Bay, USA mean that nowadays the remnant oyster population now takes a year to filter the waters of the bay. In the past this was accomplished in a week. Many areas of the southern North Sea are dominated by starfish *(Asterias rubens)*, perhaps as a result of their ability to survive trawling impacts better than most other species, and their ability to regenerate damaged body tissues. Not surprisingly then, plankton assemblages in the southern North Sea have been dominated by echinoderm larvae since the early 1980s. The extent to which disturbance of the sea bed by demersal gear might impact on the dormant eggs of certain calanoid copepods, however, appears not to have received any attention. It has been shown that the distribution of such overwintering eggs is affected strongly by sedimentation, resuspension and sediment transport caused by tidal currents.

Information on the numerical densities of eggs in sediments is scant, but the fact that they are found in the upper 1-5 cm of soft-bottom deposits means that they would be accessible to gear effects. The potential for deeper burial affecting hatching rate could provide yet another feedback between benthic fishing pressure and the composition of the overlying plankton assemblage. Similar factors might also apply to the fate of resting spores of dinoflagellates (sometimes toxic blooming species) in benthic sediments, whose hatching is affected by changing nutrient status of the surroundings (amongst other factors). Benthoplanktonic interactions would seem to offer several topics ripe for future research on fisheries impacts.

6.4 Species interactions

The role of keystone predators has been invoked to explain some well known predator-prey interactions: like lobster / sea urchin / kelp interactions in the Northwest Atlantic. However, some of these so-called trophic cascade studies are based largely on inference, and it has

6

been suggested that variability in sea urchin recruitment, juvenile mortality or physical disturbances were also regulating algal populations.

Fishing does not have consistent effects on species interactions

Sea urchins play a keystone role on some tropical coral and temperate rocky reefs and their relationships with their predators and prey are tightly coupled. However, in many marine ecosystems, predator-prey relationships are weak, transient or plastic, and the removal of one predator or prey species does not necessarily have marked impacts on the functioning of the ecosystem. Indeed, many changes in fish community structure which have been blamed on the indirect effects of fishing can be attributed to environmental change.

Changes in exploited communities are often due to losses of vulnerable species

For example the shifts between sardine or anchovy dominance in the pelagic upwelling fisheries off Peru, Southwest Africa and California would probably have occurred even had these systems not been fished. Recent studies of anchovy and sardine body scales that have accumulated in anoxic sediments beneath these upwellings suggest that the fluctuations in abundance of these fishes during the 20th century are similar to those which took place over several previous centuries and pre-dated the exploitation of these stocks.

Krill is certainly a key species in the food web of the Southern Ocean, and commercial fishing for this huge resource might be anticipated to affect populations of many marine mammals and birds that are competing for the same energy source (but see Section 5.1). However, current fishery catches are too small (<1% of stock estimates) to have large-scale effects yet. Nevertheless, there is considerable concern at the potential local effects of the persistent and substantial krill fishing within the foraging area of breeding penguins and seals around the South Shetland Islands.

The initiation of intensive fishing in an unfished system has the greatest effects on community structure

Fishing causes changes in the diversity, trophic structure and productivity of fish communities. Such changes can occur because some species suffer higher fishing mortality than others and because fishing affects feeding relationships between predator and prey fishes. Large, late-maturing species with low intrinsic rates of population increase are particularly susceptible to fishing (Figure 18). Both by-catch and target fishes with these characteristics become increasingly rare in fished systems and there is a corresponding increase in the relative abundance of small and early-maturing species. Smaller species can also experience lower fishing mortality because they are less desirable or less accessible to fishers. As a result, shifts in fish community structure result from the combined effects of differences in susceptibility and differences in fishing mortality.

Even though fishing pressure generally decreases stock abundance, it can lead to increases in production per unit biomass. Thus the growth rate of dabs (*Limanda limanda*) and plaice in the North Sea has increased, presumably because since their population density is lower and they now have access to an improved food supply (mainly of small opportunistic polychaetes that may flourish despite habitat disturbance caused by repeatedly-towed gear).

The greatest change in fish community structure occurs when fishing begins in a previously unfished ecosystem. Once systems enter a fished state, then fish diversity and overall production often remain relatively stable despite further increases in fishing intensity. For example, following low levels of poaching on a previously unexploited Seychelles marine reserve, the trophic structure and diversity of the reef fish community was typical of that in more heavily fished areas rather than that in other marine reserves (Figure 19). This suggests that studies done on systems where fishing preceded

6

Figure 18 Large slow growing species such as sharks are particularly vulnerable to high rates of fishing mortality.
(Photograph: © S. Jennings).

scientific investigations will probably not show the full extent of the effects of fishing on that ecosystem. In the North Sea, for example, records of commercial fishing activity extend further back into the nineteenth century than do our records of species' relative abundance, so we may never know the full extent of recent changes in this ecosystem.

Changes in fish community structure are often followed by changes in catches as fishers target successively less desirable species in order to maintain yields. The intensity of fishing world-wide is such that these trends are reflected in global catch statistics and the mean trophic level of fish catches has fallen since 1950 as fishers target species lower and lower down the

food chain.

Fishing-induced decreases in the abundance of voracious fish eaters, such as groupers, snappers and tunas, might be expected to lead to proliferation of their prey. However, there is little consistent evidence to suggest that this is consistently the case and prey populations are often structured by the environment, the direct effects of fishing and fluctuations in recruitment between years.

Consistent proliferation of prey may not be observed because single species can act as both predators and prey in the course of their life history and readily switch diet and feeding strategy in response to prey availability. In

Prey populations do not necessarily proliferate when their predators are caught

6

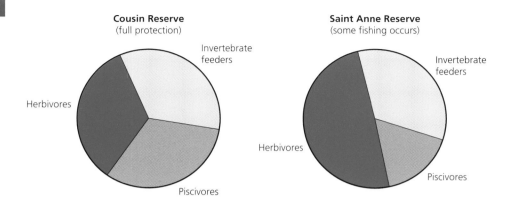

Figure 19 Structure of the fish community in two Sechelles marine reserves. The Sainte Anne Reserve is subject to some fishing and poaching and the structure of the community is similar to adjacent fished areas (From Jennings and Kaiser, 1998).

freshwater systems, however, decreases in the abundance of fished predators often lead to 'cascading' impacts on other trophic levels. Freshwater and marine systems may differ because most of the biomass in lake systems can be grouped into defined body-size categories within which the organisms tend to have a limited range of life-history traits and morphologies.

6.5 Assessing fisheries effects

Recent concerns about ecosystem effects of fishing have led scientists to 'look back in time' using long-term data series. Analyses of archived research vessel catch data from the North Sea have shown little change in overall diversity of the non-target fish component of catches, despite almost a century of high fishing effort and reductions in abundance of some species.

Long-term perspectives can be gained using a variety of other approaches. For example, episodic damage caused by fishing gear may be recorded in the growing shells of long-lived benthic bivalves, like the razor clam *(Ensis siliqua)* or the quahog *(Arctica islandica)*. If the edge of the shell becomes damaged, the shell-secretory mantle edge tissue will withdraw from the fracture and mineral particles will infiltrate the gap. Subsequent regrowth incorporates such material and a recognizable disturbance event is recorded in the shell. By sectioning shells and assessing the extent of growth arrests in shells of different ages, it is possible to build-up a historical picture of the varying impact of sea-bed disturbance due to fishing, especially in long-lived species (Figure 20).

The more recent effects of fishing on the sea bed have been studied using a variety of methods: divers, underwater television, side-scan sonar and echo-sounder devices. Side-scan sonar records from the Baltic Sea revealed that up to 35% of the sea bed was scoured by trawls in the most heavily fished areas. Using a manned submersible in the Gulf of St Lawrence, researchers some years ago estimated that 3-7% of the bottom surveyed showed evidence of disturbance.

Time-series data are needed to describe the long-term effects of fishing

Sonars, cameras and indicator species can help to detect fishing effects

6

Interpretation of such findings, however, is hampered by a lack of knowledge of the persistence of such visible tracks, which depends on sediment type and hydrographic conditions. Certainly in the Clyde Sea area, UK scallop dredge marks on maerl grounds may remain visible for many months. Equally, trawling or dredging over sheltered mud habitats may create topographic changes to the sea bed that may persist for a very long time. By contrast, in South Wales, furrows made by tractor dredgers exploiting a cockle bed with large sand ripples were barely detectable after two days. The persistence of gear tracks will also depend upon timing in relation to when the impacts and the next storm events occur to remove the evidence.

Recently, another indirect biological indicator of towed-gear damage has been proposed. The extent to which two epibenthic starfish species *(Asterias rubens* and *Astropecten irregularis)* in the Irish Sea had damaged and regenerating arms has been related to fishing effort. In heavily fished areas more than 40% of starfish may have lost arms. This particular phenomenon only has validity over the short-term (<1yr) since regenerating arms eventually become indistinguishable from undamaged ones.

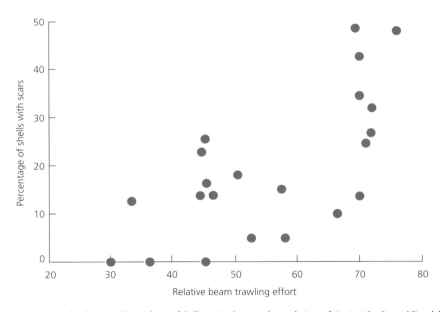

Figure 20 Relationship between the incidence of shell scars in the annual growth rings of *Arctica islandica* and Dutch beam trawling effort in the southern North Sea (data from Witbaard and Klein, 1994 and Heessen and Daan, 1996).

6

Summary

Fishing can have indirect effects on the structure of marine communities and ecosystems. Coral reefs in Kenya and the Caribbean have shifted from coral - to algal-dominated phases following fishing. However, in most cases, fishing does not have clear effects on interactions between species and any community level changes reported are usually due to the loss of vulnerable species. The greatest changes in marine communities occur when unfished areas are fished for the first time. Long-term time series data are often needed to describe fishing effects.

7 Conservation aspects and the way forward

Effective fisheries conservation is a challenge

Fishing is one of the most widespread human activities in the marine environment and, with millions of people exploiting and competing for the resource, it will always be difficult to control. Consensus among stakeholders on conservation goals is both possible and desirable, but consensus on means may be a much greater challenge. The aims of fishery management include conservation, and range from maximizing the yield of target species to ensuring that no marine mammals are killed as by-catch, or that undesirable ecosystem shifts do not occur. Given current concerns about the wider ecological impacts of fishing, there is increasing interest in developing strategies for conservation.

Poorly managed fisheries are unsustainable and have adverse impacts on marine ecosystems

This means that conserving non-target species and habitats could have higher priority than the management of a target species. In the UK, for example, general conservation legislation does not apply to commercially fished species, which are regulated (with the aim of ensuring sustainable exploitation) by separate fisheries legislation. There is, however, a proviso for Special Areas of Conservation (SACs). In an SAC, all competent authorities, including fishery authorities, are now required under the UK Conservation (Natural Habitats) Regulations 1994 (the Statutory Instrument put in place to deliver the EC 1992 Habitats Directive), to use their statutory powers to deliver the specified conservation objectives of the SAC.

7.1 Marine reserves

Marine reserves may be a useful conservation measure

Marine reserves are areas where fishing and other human activity is controlled. If all fishing is banned then the marine reserve is often known as a No-Take zone (NTZ) or closed area. NTZs can help to meet many conservation and management objectives in the environment, but they should be seen as an adjunct to existing fishery management methods. For example, a marine protected area will provide few benefits for migratory species if fishing is not controlled beyond its perimeter. NTZs should be easier to manage and police than reserves where fishing concessions are offered (though they can become magnets to poachers): NTZs protect vulnerable benthic habitats from the direct effects of fishing gears; they enhance areas of intrinsic conservation value and provide unfished control sites for the scientific study of fishing effects. Moreover, if a network of NTZs is designated, taking into account the migrations and life-histories of the inhabitants, then a series of larval sources and sinks that sustain self-perpetuating populations may be achievable, particularly for benthic species.

Reserves set up to protect discrete breeding grounds of certain migratory species can certainly be effective. The Government of Mexico has set aside Scammon's Lagoon, one of the major breeding lagoons for the Californian population of the grey whale and is regulating all boat traffic there (including whale watching) to reduce noise and propeller contacts. In 1981, the UK government closed an area within its EEZ off the Southwest coast to protect overwintering populations of juvenile mackerel (the so-called 'mackerel box', extended in 1989). A haddock nursery area was introduced in the Northwest Atlantic in 1987 to aid stock rebuilding. Designation is not, of itself, sufficient; effective enforcement must follow. The

7

ambitious Galápagos Marine Resources Reserve (GMRR) Management Plan, for example, has never been fully implemented.

The direct benefits of NTZs to fin-fish fisheries, however, are not clear-cut. As yet, there is little empirical evidence to substantiate or refute the claim that marine reserves increase fishery yield. Indeed, in the western Mediterranean Sea, mortality rates in shallow-water sea breams (*Diplodus* spp.) in NTZs did not differ significantly from those in unprotected areas, suggesting that NTZs were not necessarily a

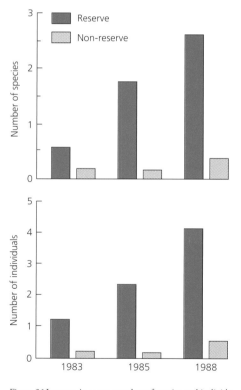

Figure 21 Increase in mean number of species and individuals of large predatory fishes in the Apo Island Reserve, Philippines after the reserve was established in 1981 (numbers per 750 m²). The species richness and abundance of large predators did not change significantly in an adjacent area open to fishing (redrawn from Russ, 1991).

sink for post-settlement fishes. It is probable that NTZs are of more relevance to the conservation of benthos than to highly migratory fish stocks in temperate waters. The question of optimal reserve size and spacing for particular species is complex and has yet to be resolved. Many small marine reserves have now been established on temperate and tropical coral reefs and, where protection from poaching has been effective, rapid increases in the abundance, size and diversity of resident fishes have resulted (Figure 21). For many of the relatively site-attached fishes on tropical reefs even small NTZs (of 1km² or less) can have appreciable conservation benefits. A network of reserves should, in theory, multiply recruitment opportunities in complex and possibly unpredictable ways. Where coastal indentation is high and /or bottom topography rugged then habitat diversity will also be high. The size and spacing of marine NTZs ought ideally to take into account the degree of habitat heterogeneity in an area.

Enforced zoning of activities, including areas where exploitation is prohibited, has been seen as providing the key for effective conservation (Figure 22). Such schemes have been successfully applied to the Great Barrier Reef Marine Park in Australia. A similar area-management scheme operates to conserve *Modiolus* grounds in Strangford Lough, Northern Ireland. Rotation of zone categorization is practised in Natal Parks, South Africa, and closed areas are rotated in the Adriatic Sea (in this case self-policed by fishers). In the past, marginal areas of rough ground could be relied upon to supply unfished refuges. However, given increasing pressures to devise methods whereby these marginal areas can be exploited it is no longer sensible to rely on *laissez faire* attitudes to conserve resources.

In rare cases, NTZs have had negative effects. Four species of *Fissurella* limpets have been shown to suffer greater trematode parasitization in a marine reserve in southern Chilean NTZ

7

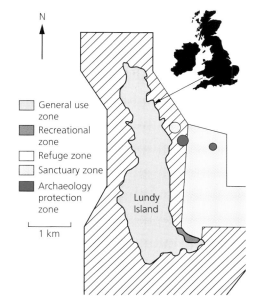

Figure 22 Zoning scheme for Lundy Island Marine Nature Reserve in the Bristol Channel, UK (modified from Gubbay, 1995).

Species-selective trawls can be designed

Trawlers can be deterred by placing obstuctions on the seabed

than in areas adjacent areas where limpets are exploited. Because the frequency of infection tends to increase with body size, this could be due to the existence of larger, older individuals inside the NTZ. But even small individuals experience greater parasitization inside the NTZ than outside it, possibly due to infection rate being related to host density. Since the parasites diminish the host's reproductive potential the limpets will be less fit where they are protected.

Researchers with the task of unambiguously demonstrating the impacts of fishing activities in the marine environment face the pragmatic problem of finding suitable pristine sites to act as control grounds for comparisons with fished areas. Nowadays it is very difficult to find sea beds that have remained untouched by fishing activity in shelf seas, though NTZs could provide recovering sites. As the above Chilean example shows, the absence of fishing may not be the only difference between NTZs and fished

areas, so comparisons must be made with caution.

7.2 Practical conservation measures

Some quite simple practical expedients will often assist conservation requirements. Experiments are in progress to find effective ways to increase the acoustic detectability of fishing nets to deter the entrapment of cetaceans, without compromising their efficiency with respect to target species. Using biodegradable clips to secure the access ports would ensure that pots would cease fishing after time if lost. The possibility that fishing nets might be made of biodegradable materials is also a topic of current research.

The problem of 'ghost' fishing by lost, bottom-snagged nets, is being addressed in certain fishery areas by regular bottom cleaning using grapples. The use of transponders on nets might also facilitate their retrieval if lost. Since the best fishing is often around the margins of hard ground or wrecks, it is often the most productive areas that are the most likely to accumulate netting entanglements. Trawling is often blamed for the loss of static gears on grounds exploited by both methods, creating vocal antipathy between trawl- and pot-fishers. Large numbers of huge irregular concrete blocks have been deployed to deter trawling activity, particularly in Spanish mediterranean seagrass habitats (Figure 23). Habitat restoration measures are now in place in many countries to counter loss of seagrass meadows, kelp forest, mangrove and coral reef habitats, ideally as part of Integrated Coastal Zone Management (ICZM) schemes.

The development of methods that reduce excess catch, or fish more selectively, is one way of reducing incidental fisheries mortality. Prompted by the Cornish mackerel problem, described in Section 6.2, trawler skippers

7

devised a method of reducing excess catches by cutting a 'swilly hole' or 'zipper' 2-3m long, positioned in the top of the cod end such that it could retain no more than the vessel could carry or process in a day, and through which the excess catch could swim.

By-catch reduction devices (BRDs), like the fish eye, a metal support sewn into the codend

(a)

(b)

Figure 23 Concrete antitrawling blocks (a) awaiting deployment, and (b) deployed *in situ* in seagrass *(Posidonia)* beds off the Spanish Mediterranean coast (Photograph: © A.A.Ramos-Esplá)

7

Turtle exclusion devices keep turtles out of shrimp nets

Square mesh panels in trawl nets allow undersized fish to escape

Modifications to fishing practices can reduce by-catch

of US shrimp trawls to let finfish escape, can reduce by-catch by 50%, but losses of shrimp mean that it is unpopular with shrimp fishers. The selectivity of otter trawls may also be modified by varying the size and shape of the mesh panels used in different parts of the net and the circumference and length of the cod-end. Separation of species captured (e.g soft-bodied roundfish from spiky Norway lobster), will minimize damage caused by their coming into contact, and can be achieved by inserting a horizontal panel across the net bag. Norway lobsters never rise more than 70cm from the sea bed, whereas small roundfish will tend to rise as they tire and pass into the upper zone of the separator net. The upper codend can then have a larger mesh than the lower. Similar large (square)-mesh top panels have been shown to be effective in facilitating escape of whiting *(Merlangius merlangus)* and cod from beam trawls equipped with tickler chains targetting flatfish. However, their efficiency in chain-mat beam trawls depended on the scale of the gear and size of the vessel, since round fishneed enough time to swim upwards once past the mouth of the gear.

The use of a single top panel in mid-water trawls, instead of two overlapping layers of netting, has been shown to reduce by 75% the by-catch of undersized walleye (Alaskan) pollock *(Theragra chalcogramma)* by facilitating the escape of juveniles. This is a significant development, in the largest trawl fishery in the USA. In UK waters, the insertion of square mesh panels into nephrops trawls has reduced the by-catch of juvenile whiting. Square mesh panels reduce by-catch because square meshes remain open under tension, allowing better egress of unwanted organisms, whereas conventional diamond-shaped meshes pull shut.

Recent work in the Northeast Mediterranean,

comparing the effect of trawl codend mesh size and shape on discards, has shown that the 14mm diamond-shaped mesh presently in use is more harmful (catching more non-commercial and fewer commercial species) than the 20mm diamond or 20mm square mesh codends, which allowed significantly more individuals and species to escape capture. Selectivity is not simply a mechanical sieving process. The ability to escape is determined by body condition or swimming capacity as well as by the relationship between cod-end girth and mesh opening, and so may well vary seasonally.

Other modifications to fishing gears that mitigate by-catch of unwanted or endangered species have been developed. Turtle exclusion devices (TEDs) - metal grids with an escape hatch - are inserted across the codends of prawn trawls to reduce fish by-catch and eliminate capture of turtles. These are now used widely in the USA and Australia. The US Court of International Trade recently imposed a unilateral ban on the importation of prawns from 52 countries that do not have specific turtle conservation programmes, creating a furore both at home and abroad (e.g. in India) among shrimp fishers. The Texas shrimp association, for instance, claimed that TEDs cut shrimp catches by 20%, although federal records indicated that 5% was a more realistic figure. Excluder panels have also been designed for pelagic trawls to reduce cetacean by-catches. These consist of a panel of ropes attached to the bottom panel directly behind the footrope, guiding cetaceans upwards (bottlenose dolphins proved to be reluctant to escape downwards) to escape through enlarged meshes in the top panel.

Great progress has been made in reducing dolphin mortality in purse seines used to capture

7

tunas. The effective measures included requiring skippers to perform the backdown procedure (Section 5.2), modifications to the gear such as the incorporation of a panel of fine mesh set at the apex of the backdown channel, the use of skiffs or speedboats to prevent collapse of seines and the use of rafts or swimmers to rescue dolphins remaining in the net after backdown. Perhaps more importantly, coordinated international education efforts have raised the profile of the dolphin by-catch problem (see section 5.2).

Fishing technology is now so efficient that one catch can exceed the hold capacity of a fishing vessel. It is not unusual for the catch of a big purse seine to be too heavy to lift bodily from the water (40-50 tonnes on one set), and captured fish must then be brailed using a powered dip net or pumped on-board instead. When catches exceeded the hold capacity of a purse seiner, excess fish used to be held alive in the submerged purse and released by 'slipping' the end of the purse free from the ship. However, UK research on mackerel in 1976 showed that, although they swam away, they did so only to die, having suffered irreversible damage to the skin's osmoregulatory ability in the close confines of the net. In extreme cases, they developed blisters which burst, leaving areas of unprotected flesh.

Financial pressures can cause regulations to be ignored

Financial pressures on fishers can result in environmentally sympathetic regulations being ignored. For example, it is noteworthy that Alaskan demersal longline fishers reported three short-tailed albatrosses killed in 1995. Only about 100 or so pairs of this endangered species now breed at two sites off Japan (see Section 5.1). Since notifying members that the fishery could be closed if any more of this seabird should be killed on longlines, none has been reported. It is unknown whether this is due to improved mitigation measures or astute

Effective enforcements of fishery regulations is essential

lack of reporting. Instances have also come to light of captured turtles being slashed deliberately to give the appearance that they had succumbed to shark attack and not perished from net entanglement. By-catch sharks are often dumped back into the sea alive after having their fins cut off for sale in Far Eastern markets. Shark finning is illegal in the US Atlantic shark fishery but remains legal in the Pacific.

Substantial reductions in net-mortality of some non-target species could be achieved with sensible action based on current ecological knowledge. Studies indicate that pelagic species most at risk from net entanglement are those marine predators that pursue their prey underwater (like divers, penguins, shearwaters, guillemots, seals, dolphins, porpoises and sharks) and aggregate in dense foraging groups as do colonially-nesting gannets and penguins. Greatest net-catches occur during periods when prey become available in areas frequented by predators and fishers. Exclusion of fishing (sometimes in total, sometimes with particular types of gear) from sensitive areas has to be taken seriously as a means of protecting vulnerable species, if only at specific times of year. For example, stopping prawn trawling near rookery beaches during peak turtle nesting periods has been effective in reducing the mortality of female loggerhead turtles in Australia. The protection of vulnerable habitats, for instance, will provide nursery areas for fish and shellfish that can be captured later outside their perimeter. In New Zealand, in 1997, the Government closed early the lucrative squid fishery around Auckland Island when the by-catch (100) of Hooker's sea lion *(Phocarctos hookeri)*, a threatened species with a world population of only 13,500, exceeded the limit allowed under New Zealand legislation (73). In November 1996, the US Northeast Atlantic drift gillnet was closed on an emergency basis because of the

7

threat posed to the northern right whale *(Eubalaena glacialis)*, the world's most endangered large whale species. It remains closed. The use of by-catch limits to enforce fisheries closure, or movement to new areas, can only work with scientific observers aboard all vessels; otherwise unsupervised fishers will be reluctant to register the true by-catch of endangered species, when such data may result in their exclusion from a lucrative fishery.

In the USA, new legislation requires the use of a deliberately simple formula to determine a mortality limit for marine mammal by-catch in a fishery. If the estimated by-catch is greater than this limit, then a Take Reduction Team is formed composed of fishermen, scientists, managers and environmentalists. The Team must agree on measures to reduce mortality to below the calculated limit. Implementation of these measures has resulted in gear restrictions and time-area closures in bottom-set gillnet fisheries to protect harbour porpoises.

In recent years, a number of experimental trials in waters off the Northeast USA and in the North Sea have used active acoustic deterrents (pingers) in attempts to reduce porpoise by-catch in bottom-set gillnets. These have been highly successful, demonstrating a by-catch reduction of up to 90%. Pingers have also been shown to reduce cetacean by-catch by 80% in the swordfish fishery of fthe West coast of the USA. Trials are being conducted in 1998/99 in the Celtic Sea to determine whether pingers could be used routinely by a commercial fishery as part of mitigation measures.

A number of devices have been developed to reduce fishery losses to seals. Acoustic harassment devices (seal scrammers) have been deployed to some effect around aquaculture facilities, but their use is limited by their high energy requirements and cost. Cape fur seals in South Africa have been shown to respond to playback of killer whale *(Orcinus orca)* sounds, but without being caused to flee. They did move away from weighted firecrackers exploded underwater and 0.303 rifle bullets fired into the water, and this resulted in the development of an arc-discharge transducer which developed similar underwater compression waves. However, this apparatus had equivocal effects on seal behaviour at fishing gear and, one may presume, unwanted side-effects in scaring fishes too.

Mariculture and/or extensive high-seas ranching have been suggested, and in some regions attempted (recently even for tuna by the Japanese), to compensate for the impact of exploitation on wild stocks. In all cases involving large-scale cultivation, however, there have been consequences for the natural environment, the genetic structure of the cultivated population, transmission of disease and parasites, and impact on wild counterparts. This suggests that such approaches should not automatically be regarded as providing the panacea for countering one impact of capture fisheries on the marine environment. It is noteworthy though that in certain small-scale operations not involving fish production for food, e.g. seahorse conservation, bioremediation measures may well usefully include artificial cultivation.

Appropriate enforcement of fishing activities within Exclusive Economic Zones (EEZs) reaching up to 200 miles offshore are potentially the single most important development facilitating sustainable exploitation of fish stocks. Although 99% of marine fisheries catch is currently taken within such limits, any one nation's catch will include fish taken from outside home waters (25% of EC landings) due to joint-venture agreements between countries.

7

It is now possible for the whereabouts of all fishing boats to be monitored by satellite to ensure that such international boundaries to fishing and closed areas are respected. Such initiatives are already in progress in North America, in Europe and in the Southern Ocean. Traditionally, fleet monitoring was accomplished by fishery patrol vessels and aircraft surveillance and even, in some countries, by undercover agents working as crew. Unsurprisingly the most effective enforcement of closed areas is likely to be near militarily sensitive sites within the jurisdiction of armed forces. Fishery regulation by 'gunboat diplomacy' occurred most recently during the Canadian 'turbot war' of 1995.

Straddling stocks that migrate across national boundaries are a shared resource

While helping with assessment of commercial stocks, however, EEZs have not yet addressed the problems of open access to resources within national boundaries. Straddling stocks, those which cross boundaries between EEZs, represent an additional complication. The recent (1995) UN Convention on Straddling Stocks resulted from the recognition that lack of regulation in traditional and developing high-seas fisheries is in no-one's best interest. The introduction of closed areas outside EEZs will require a culture change in Governments and fisheries regulation authorities. This development is still some way off. Of relevance in this connection is the Southern Ocean experience with albatross by-catches; it is not only necessary to protect species within the CCAMLR area (birds' breeding areas), but also to reduce death rates of the same species breeding within, but captured outside, that area and *vice versa*. Southern Ocean albatrosses are probably at greater risk outside the CCAMLR area than within it.

The idea that fisheries should be closed when by-catches quotas are exceeded (Section 5.2) is controversial. In practical terms this can only be operated through independent international scientific observers on fishing vessels. In 1995, the CCAMLR-IMALF group emphasized that data from scientific observers were crucial to accurate reporting of incidental mortalities in regulated Southern Ocean longline fisheries.

It is deceptively easy to say "Ban all discarding". How would this be enforced? No such ruling can ever prejudice a skipper's responsibility to discard any catch that might jeopardize the safety of the vessel, but this clearly provides a convenient uncheckable excuse for discarding. Patently, scientific observers cannot sail on all fishing vessels. There is, however, one important model of ecosystem management to look to for inspiration, that provided by the CCAMLR agreement (signed in 1981), the far-reaching conservation treaty which sets out to manage the whole Southern Ocean ecosystem for rational use.

It is encouraging to see that many practical conservation measures are being considered now by those concerned with the management of marine fisheries. In November 1995, the FAO Council adopted the final text of the Code of Conduct for Responsible Fishing, which has been ratified by most member states. The Code includes advice on the use of selective fishing gear and on the protection of the aquatic environment. Although the Code is voluntary, it has one legally binding component: the agreement 'to promote compliance with international conservation and management measures by fishing vessels on the high seas'.

Another approach to ensuring the rational exploitation of marine ecosystems has been through mobilizing consumer concern for the marine environment. For example, the Marine Stewardship Council (MSC), an independent world council spearheaded by WWF and Unilever, was set up in 1997 with the aim 'to ensure the long-term viability of global fish pop-

7

ulations and the health of the marine ecosystems on which they depend'. Their intention is to accredit firms which can certify that particular voluntarily-nominated fisheries have been carried out according to MSC principles and criteria, and to signify such by the use of product labelling. While the MSC scheme is an interesting development, it is difficult to see how consistent criteria could be applied worldwide, and such schemes may falter due to the inadequacy of scientific data for many fisheries.

Increasing public awareness of the environmental impacts of fishing might well result in a shift in the perception of fishers from dignified seafarers to pillagers of the environment, i.e. fishers may soon suffer the hostility that farmers have increasingly faced in recent years. Political pressures can then build to the extent that certain fisheries might face closure regardless of scientific arguments. Certainly, the ethical concerns causing US buyers to boycott tuna caught in dolphin-unfriendly ways and which led to "dolphin safe" labelling of tuna products in 1991, has shifted the US tuna fishery from setting purse seine nets around dolphins to setting on free-swimming schools of tuna or on floating objects, like logs.

The problem with tuna fishing remains since it is significantly more successful when dolphins are present. It is the tuna, particularly the larger mature yellowfin, that follow the dolphins, possibly taking advantage of the cetaceans' superior fish-finding abilities. Paradoxically, with appropriate mitigation methods in place (see above), setting purse seines on dolphins and accepting some dolphin mortality probably remains the ecologically-, fishery- and economic-best practice available for tuna exploitation, given the large by-catches of other fish with other setting methods (Section 5.2). But in the USA, where the issue has assumed great symbolism, it having been the driving force behind the enactment of the Marine Mammal Protection Act in 1972, environmental groups continue to press for a total ban on dolphin sets.

Much clearly remains to be done to regulate global fisheries in an environmentally sensitive and ethical manner, but with honesty and integrity on all sides and a willingness to learn from past mistakes, answers should emerge to the global problems of the conservation of fished stocks, the integrity of the habitats that support them, and the future of fishers who exploit them legitimately for their livelihood.

Summary

Effective fisheries conservation will maximize the long-term yields of protein and income from fisheries while seeking to minimize environmental impacts. Marine reserves help to protect vulnerable habitats or species of conservation concern and to provide 'control' sites for scientific studies of fishing effects. Many fishing gears and fishing methods can be modified to reduce by-catches and these modifications are increasingly required by law.

FURTHER READING

Allen, R.L. (1985). Dolphins and the purse-seine fishery for yellowfin tuna. *Marine mammals and fisheries,* (Ed. by J.R. Beddington, R.J.H. Beverton & D.M. Lavigne), pp. 236-252. George Allen and Unwin, London.

Alverson, D.L., Freeberg, M.H., Murawski, S.A. & Pope, J.G. (1994). *A global assessment of fisheries by-catch and discards.* Food & Agriculture Organization, Rome.

Alexander, K., Robertson, G. & Gales, R. (1997). The incidental mortality of albatrosses in longline fisheries. *Report on the Workshop from the First International Conference on the Biology and Conservation of Albatrosses, Hobart, Australia.* Australian Antarctic Division, Tasmania.

Britton, J.C. & Morton, B. (1994). Marine carrion and scavengers. *Oceanography & Marine Biology, an Annual Review,* **32,** 369-434.

Bohnsack, J.A. (1998). Application of marine reserves to reef fisheries management. *Australian Journal of Ecology, 23,* 298-304.

Camphuysen, C.J., Ensor, K., Furness, R.W., Garthe, S., Huppop, O., Leaper, G., Offringa, H. & Tasker, M.L. (1993). *Seabirds feeding on discards in winter in the North Sea.* Netherlands Institute of Sea Research, Den Burg, Texel.

Churchill, J.H. (1989). The effect of commercial trawling on sediment resuspension and transport over the Middle Atlantic Bight continental shelf. *Continental Shelf Research, 9,* 841-864.

Croxall, J.P., Rodwell, S. & Boyd, I.L. (1990). Entanglement in man-made debris of Antarctic fur seals at Bird Island, South Georgia. *Marine Mammal Science,* **6***,* 221-233.

Dayton, P.K., Thrush, S.F., Agardy, M.T. & Hofman, R.J. (1995). Environmental effects of marine fishing. *Aquatic Conservation: Marine and Freshwater Ecosystems,* **5***,* 205-232.

Dye, A.H., Branch, G.M., Castilla, J.C. & Bennett, B.A. (1994). Biological options for the management of the exploitation of inter-tidal and subtidal resources. *Rocky shores: exploitation in Chile and South Africa* (Ed. by W.R.Siegfried). Springer-Verlag, Berlin.

FAO (1991). *Environment and the sustainability of fisheries.* Food & Agriculture Organization, Rome.

FAO (1995). *World Fishery Production 1950-1993.* Food & Agriculture Organization, Rome.

Furness, R.W. (1982). Competition between fisheries and seabird communities. *Advances in Marine Biology,* **20***,* 225-327.

Furness, R.W. (1992). *Implications of changes in net mesh size, fishing effort and minimum landing size regulations in the North Sea for seabird populations.* Joint Nature Conservaton Committee Report 133, Peterborough.

Gray, J.S. (1997). Marine biodiversity: patterns, threats and conservation needs. *Biodiversity & Conservation,* **6***,* 153-175.

Gubbay, S. (1995) *Marine protected areas: prin-*

8

ciples and techniques for management. Chapman and Hall, London.

Hall, M.A. (1998). An ecological view of the tuna-dolphin problem: impacts and trade offs. *Reviews in Fish Bilogy and Fisheries*, **8**, 1-34.

Hall, S.J. (1999). *The Effects of Fishing on Marine Ecosystems and Communication*. Blackwell Science, Oxford.

Heessen, H.J.L. & Daan, N. (1996). Long-term changes in ten non-target North Sea fish species. *ICES Journal of Marine Science,* **53**, 1063-1078.

Hudson, E. & Mace, G. (Eds) (1996). *Marine fish and the IUCN Red List of threatened animals*. Zoological Society, London.

ICES. (1995). Report of the study group on ecosystem effects of fishing activity. *International Council for the Exploration of the Sea, Cooperative Research Report,* 200, 1-120.

Jennings, S. & Polunin, N.V.C. (1996). Impacts of fishing on tropical reef ecosystems. *Ambio*, **25**, 44-49.

Jennings, S. & Kaiser, M.J. (1998). The effects of fishing on marine ecosystems. *Advances in Marine Biology*, **34**, 201-351.

Kaiser, M.J. & Spencer, B.E. (1994). Fish scavenging behaviour in recently trawled areas. *Marine Ecology Progress Series*, **112**, 41-49.

Kaiser, M.J. & de Groot S.J. (eds) (2000). *The Effects of Fishing on Non-Target Species and Habitats*. Blackwell Science, Oxford.

Kaiser, M.J., Bullimore, B., Newman, P., Lock, K. & Gilbert, S. (1996). Catches in `ghost fishing' set nets. *Marine Ecology Progress Series*, **145**, 11-16.

King, M. (1995). *Fisheries biology: assessment and management*. Fishing News Books, Oxford.

Lindeboom, H.J. & de Groot, S.J. (Eds) (1998). *The effects of different types of fisheries on the North Sea and Irish Sea benthic ecosystems*. NIOZ-Rapport 1998-1, RIVO-DLO Report C003/98 Netherlands Institute for Sea Research, Texel.

Macpherson, E., Biagi, F., Francour, P., Garcia-Rubies, A., Harmelin, J., Harmelin-Vivien, M., Jouvenel, J.Y., Planes, S., Vigliola, L. & Tunesi, L. (1997). Mortality of juvenile fishes of the genus *Diplodus* in protected and unprotected areas in the western Mediterranean Sea. *Marine Ecology Progress Series*, **160**, 135-147.

McClanahan, T.R., Kamukuru, A.T., Muthiga, N.A., Gilagabher Yebio, M. & Obura, D. (1996). Effect of sea urchin reductions on algae, coral, and fish populations. *Conservation Biology*, **10**, 136-154.

Moore, P.G. & Howarth, J. (1996). Foraging by marine scavengers: effects of relatedness, bait damage and hunger. *Journal of Sea Research,* **36**, 267-273.

OECD. (1997). *Towards sustainable fisheries: economic aspects of the management of living marine resources*. Organization for Economic Co-operation and Development, Paris.

Pauly, D. & Christensen, V. (1995). Primary production required to sustain global fisheries. *Nature*, **374**, 255-257.

Pauly, D., Christensen, V., Dalsgaard, J., Froese, R. & Torres, F. (1998). Fishing down marine food webs. *Science,* **279**, 860-863.

Perrin, W.F., Donovan, G.P. & Barlow, J. (1994). *Gillnets and cetaceans. Reports of the International Whaling Commission* (Special Issue **15**), 629pp.

8

Polunin, N.V.C. & Roberts, C.M. (Eds) (1996). *Reef fisheries*. Chapman & Hall, London.

Probert, P.K., McKnight, D.G. & Grove, S.L. (1997). Benthic invertebrate by-catch from a deep-water trawl fishery, Chatham Rise, New Zealand. *Aquatic Conservation: Marine and Freshwater Ecosystems*, 7, 27-40.

Ramsay, K., Kaiser, M.J., Moore, P.G. & Hughes, R.N. (1997). Consumption of fisheries discards by benthic scavengers: utilization of energy subsidies in different marine habitats. *Journal of Animal Ecology*, **66**, 884-896.

Rogers, S.I., Maxwell, D., Rijnsdorp, A.D., Damm, U. & Vanhee, W. (1999). Fishing effects in northeast Atlantic shelf seas: patterns in fishing effort, diversity and community structure. IV. Can comparisons of species diversity by used to assess human impacts on demersal fish faunas *Fisheries Research*, **40**, 135-152.

Rowley, R.J. (1994). Marine reserves in fisheries management. *Aquatic Conservation: Marine and Freshwater Ecosystems*, 4, 233-254.

Russ, G. R. (1991) Coral reef fisheries: effects and yields. *The ecology of fishes on coral reefs*, (Ed. by P. F. Sale), pp. 601-635. Academic Press, San Diego.

Sainsbury, J.C. (1986). *Commercial fishing methods: an introduction to vessels and gear*. Fishing News Books, Oxford.

Sale, P.F. (Ed.) 1991. *The ecology of fishes on coral reefs,*. Academic Press, San Diego.

Wade, P.R. (1998). Calculating limits to the allowable human-caused mortality of cetaceans and pinnipeds. *Marine Mammal Science,* **14**, 1-37.

Witbaard, R. & Klein, R. (1994). Long-term trends on the effects of the southern North Sea beamtrawl fishery on the bivalve mollusc *Arctica islandica* L. (Mollusca, Bivalvia). *ICES Journal of Marine Science*, **51**, 99-105.

Anoxia (anoxic) - the absence of oxygen.

Backdown procedure - manoeuvre in which a purse seining vessel reverses to collapse of the rim of the net and allow trapped dolphins to escape.

Benthos (benthic) - organisms inhabiting the sea bed.

Berm - a linear ridge of sediment.

Biomass - the total mass of organisms of specified species in a specified area.

Bioremediation - the use of living organisms to restore impacted habitats.

Bioturbation - the mixing of sediment caused by burrowing, feeding or other activity of living organisms.

Brailing - transfer of catch from the net to a fishing vessel (usually a purse seiner), by means of dip netting (usually mechanical).

By-catch - that part of the catch not composed of target species.

Byssate - the production of byssus threads by certain bivalve molluscs. Byssus threads anchor the bivalve to the substratum.

Cetacean - marine mammals of Order Cetacea (whales, dolphins and porpoises)

Cnidarian - organisms of phylum Cnidaria (jellyfish, corals and sea anemones).

Codend - the rear end of the trawl where catch accumulates.

Conservation - the protection, maintenance and rehabilitation of native biota, their habitats and life support systems to ensure ecosystem sustainability and biodiversity.

Continental shelf - submerged area landwards of the 200m isobath (legal definition).

Continental slope - submerged area seaward of the 200m isobath, but landwards of the continental rise.

Copepod - member of a Class of small aquatic crustaceans.

Creel - box-shaped or rectangular mesh-covered trap with entrance tunnels designed to catch shellfish.

Deep sea - a general term for those regions lying beyond the shelf break (depths >200m).

Demersal - those fish species or fishing gears which live or operate close to the sea bed.

Diel - daily.

Discarded catch (discards) - that proportion of the catch returned to the sea for economic, legal or personal reasons.

Discard mortality - discard mortality rate multiplied by discarded catch.

Discard mortality rate - the proportion of the discarded catch that dies as a result of the catching or handling process.

Drive netting - a technique of fishing involving driving fish into fixed nets or traps, using sound or vibration.

Eelgrass- aquatic seed-producing flowering marine plant (angiosperm).

Epibenthic - living on or immediately above the sea bed.

Extirpation - the loss of a local population as distinct from an entire species (extinction).

Fishing effort - the amount of time a given combination of inputs (such as the vessel and gear) spends searching for or catching fished resources.

Fleet - 1. a collective name for a linked sequence of nets or pots, or 2. a group of vessels active in a specified fishery.

Floc - an aggregation of fine particles such as settled organic detritus, faecal pellets and fine inorganic sediment particles; a coacervate.

'Ghost' fishing - the continued fishing of gears (nets, traps) lost by fishers.

Gill net - a loosely set and near invisible wall of fine netting that traps fish by the gill covers.

Hermatypic - reef-forming corals that contain symbiotic algae within the polyps.

High grading - splitting the catch of a single species into high and low value components, and disposing of the low value catch in order

9

to maximize profits from a legal quota.

Incidental catch - those species which are caught infrequently and have some commercial value, they are neither target nor discard species.

Iteroparity - organisms having repeated reproductive cycles (cf. semelparous)

Keystone species - a species having a major influence (as predator or prey) upon the abundance of other species in a community.

Krill - from the Norwegian for the euphausiid *Euphausia superba* the principal food of the baleen whales, and many other species, in the Southern Ocean.

Licence - A licence (or permit) is a document giving the right to fish according to terms established by the regulatory authority.

Litter - any discarded or lost items of fishing gear (or fragments thereof).

Monofilament net - a net made from fine nylon monofilament that is almost transparent in water.

Nekton (nektonic) - organisms living in the water column that are capable of significant independent locomotion.

Nematocyst - the stinging cell that is the characteristic feature of the phylum Cnidaria (or Coelenterata).

Niche - the ecological role of a species in a community.

Otolith - bones in the inner ear of fishes. When sectioned they show annual banding patterns that can be used to age fish.

Overfishing - fishing with a sufficiently high intensity to reduce the catch rates that a fish population should be capable of sustaining.

Pelagic - pertaining to the water column.

Phylum (pl.phyla) hence phyletic - the rank within the zoological hierarchy of classification below Kingdom and above Class.

Plankton - organisms of the water column that are incapable of significant independent locomotion relative to the strength of currents

Planulae - free-swimming larvae of cnidarians.

Pot - hemispherical or barrel-shaped trap. The former tend to be made of wicker work with a funnel at the top (ink pot), the latter type may be of horizontal slats or wire mesh.

Protandr (y-ous)- an organism in which male and female reproductive organs co-occur (hermaphrodite) that assumes a functional male condition during development before reversal to a functional female state (cf. protogynous).

Protogyn (y-ous)- a hermaphrodite that functions sexually initially as a female and then changes sex to male (cf. protandrous).

Pseudofaeces - particulate matter selectively sorted by and rejected from the gills of bivalve molluscs without ingestion into the gut.

Purse seine - an encirling net with a base that sinks rapidly and a headline that floats. Used to encircle shoals of fish before the base is drawn in (pursed) to trap the fish.

Rock hopper rig - a modified otter trawl with metal rollers or balls (bobbins) attached to the ground rope. These help the net to keep clear of, and run over, obstacles in the path of the net.

Roundfish - fishes that are rounded in transverse section (as opposed to flatfish).

Semelparity - organisms that reproduce once in their lifetime (cf. iteroparity)

Shooting - laying out of line or net.

Southern Ocean - nondefinitive term for the portions of the Atlantic, Pacific, and Indian Oceans surrounding Antarctica; often more specifically applied to the Antarctic waters south of the Antarctic convergence (Polar Front).

Straddling stocks - fish stocks that migrate between EEZs.

Stenotopic - organisms tolerant of a narrow range of habitat conditions.

Styrofoam -expanded polystyrene packaging material.

Sweep - the rope (usually wire) between the trawl otter board and net.

Target catch - species or species assemblages

9

that are primarily sought in a fishery.

Thallus (pl. - thalli) - the plant body of an alga.

Thermocline - an abrupt discontinuity in the temperature profile of a stratified water column with warmer water overlying denser colder water.

Tickler chain - a chain, or chains, stretching between the shoes of a beam trawl to disturb the surface of the sediment in front of the mouth of the net.

Trammel net - a bottom-fished net made from three sheets of netting joined, top and base, to lay together as one wall. The two outer sheets are of larger mesh than the loosely hung inner sheet. When fished, the net wall hangs vertically. If a fish strikes the wall its propulsive force carryies the smaller mesh through the larger mesh, trapping and entangling the fish in a net pocket.

Trash - a colloquial term for material discarded by fishers.

Transponder - an electrical-transmitting device.

Trematode - a member of the Trematoda, a Class of exclusively parasitic flatworms.

Trolling - a fishing method that involves towing a lure or bait behind a moving vessel.

Trophic - 1. nutrition, or 2. steps in a food chain (levels).

Tunicate - a member of the subphylum Tunicata, or sea squirts, a group of solitary or colonial free-living marine chordates.

Turbidity current - bottom current generated by water of very high density and caused by excessive turbidity, which flows down a sloping bottom (e.g. underwater canyon) at high speed.

Upwelling - when deep nutrient-rich waters are brought to the surface as a result of offshore movement of wind-driven surface waters. Upwelling regions have especially high productivity.

Yield - harvest of target species from a body of water.

10 ACRONYMS

ASCOBANS Agreement on Small Cetaceans of the Baltic and North Seas
BRD By-catch Reduction Device
CCAMLR Convention on the Conservation of Antarctic Marine Living Resources
DDT Ethane,1,1,1-trichloro-2,2-bis (p-chlorophenyl)-
EC European Commission
EEZ Exclusive Economic Zone
EU European Union
FAD Fish Aggregating Device
FAO Food & Agriculture Organization of the United Nations
GATT General Agreement on Tariffs and Trade
IATTC Inter-American Tropical Tuna Commission
ICES International Council for the Exploration of the Sea
ICZM Integrated Coastal Zone Management
IMALF Working group on the Incidental Mortality Associated with Longline Fishing

IUCN International Union for the Conservation of Nature
IWC International Whaling Commission
MARPOL Marine Convention for the Prevention of Pollution from Ships
MSC Marine Stewardship Council
mt Million tonnes
my BP Million years before present
NTZ No-Take zone
OECD Organization for Economic Co-operation and Development
SAC Special Area of Conservation
SCUBA Self-contained underwater breathing apparatus
SPR Species population region
t metric tonne
TAC Total allowable catch
TED Turtle exclusion device
UK United Kingdom
UN United Nations
USA United States of America
WWF Worldwide Fund for Nature

Professor J.P.Croxall, British Antarctic Survey, Natural Environment Research Council, High Cross, Madingley Rd, Cambridge, England CB3 0ET [e-mail j.croxall@bas.ac.uk]

Professor R.W.Furness, Institute of Biomedical & Life Sciences, Division of Environmental & Evolutionary Biology, Graham Kerr Building, University of Glasgow, Glasgow, Scotland G12 8QQ [e-mail gbza16@udcf.gla.ac.uk]

Dr P.S.Hammond, NERC Sea Mammal Research Unit, The Gatty Marine Laboratory, School of Environmental and Evolutionary Biology, University of St Andrews, St Andrews, Fife, Scotland KY16 8LB [e-mail p.hammond@smru.st-andrews.ac.uk]

Dr S. Jennings, formerly at the School of Biological Sciences, University of East Anglia, Norwich, England NR4 7TJ. Present address: The Centre for Environment, Fisheries & Aquaculture Science Laboratory, Pakefield Rd, Lowestoft, Suffolk, England NR33 0HT [e-mail S.Jennings@cefas.co.uk]

Dr M.J.Kaiser, School of Ocean Sciences, University of Wales-Bangor, Menai Bridge, Anglesey, Gwynedd, North Wales LL59 5EY [e-mail m.j.kaiser@sos.bangor.ac.uk]

Dr E.Macpherson, Centro de Estudios Avanzados de Blanes (C.S.I.C.), Cami de Santa Bàrbara s/n, 17300 Blanes, Girona, Spain. [e-mail macpherson@ceab.csic.es]

Professor P.G.Moore, University Marine Biological Station Millport, Isle of Cumbrae, Scotland KA28 0EG [e-mail pmoore@udcf.gla.ac.uk]

Dr S.I. Rogers, The Centre for Environment, Fisheries & Aquaculture Science Laboratory, Pakefield Rd, Lowestoft, Suffolk, England NR33 0HT [e-mail S.I.rogers@cefas.co.uk]